T0203911

Introduction to GEOPHYSICAL FORMATION EVALUATION

James K. Hallenburg

CRC Press
Taylor & Francis Group
Boca Raton London New York

CRC Press is an imprint of the
Taylor & Francis Group, an **informa** business

CRC Press
Taylor & Francis Group
6000 Broken Sound Parkway NW, Suite 300
Boca Raton, FL 33487-2742

First issued in paperback 2019

© 1998 by Taylor & Francis Group, LLC
CRC Press is an imprint of Taylor & Francis Group, an Informa business

No claim to original U.S. Government works

ISBN-13: 978-1-56670-263-8 (hbk)
ISBN-13: 978-0-367-40075-0 (pbk)

**Visit the Taylor & Francis Web site at
http://www.taylorandfrancis.com**

**and the CRC Press Web site at
http://www.crcpress.com**

Contents

Preface

This volume, *Introduction to Geophysical Formation Evaluation*, is intended as a practical volume for the engineer, environmentalist, geologist, miner, exploration manager, and student. Although it contains some relatively detailed analyses, it is not intended to be used to design geophysical instruments, but rather, how to use them. Its focus is on the petroleum usage of geophysical methods: petroleum-funded research and engineering is where about 90% of present-day geophysical methods originate. Non-hydrocarbon methods and practice, as well as other-than-borehole methods are planned for subsequent books. It is important to know where these methods originate.

Some of the newer instruments and/or techniques have purposely been left out; others have been included, even though they are not yet available in the field. Still others have also been included, although they have fallen into apparent disuse. The author has attemptted to judge the future value of all of the instruments and techniques; obviously, this attempt is not going to be 100% successful, and revisions will no doubt have to be made to correct these mistakes. This is a rapidly changing field; even the approaches and character of the discipline are changing, making it difficult to keep up them.

There are many books currently available on how to interpret logs, how to design instruments, and how to conduct exploration. *Introduction to Geophysical Formation Evaluation* attempts to answer why a certain procedure is used and why a particular instrument is chosen over another.

A chapter on the history of logging and formation evaluation has been included. It appears to be important, as well as interesting. Perhaps some of the myths and errors of word-of-mouth stories can be corrected. For example, the author was the Project Engineer for the MicroLog and is continually amazed at how many "were present for the first test of the MicroLog". There are several versions of "the first Schlumberger log." Frankly, I do not know which is the first one. The one shown in this text is certainly one of the early ones.

The references combine both the sources of some of the information included in the text and additional books and papers for further investigation by the reader.

A note of warning!! Numerous tables have been included in this volume. Bear in mind that the values listed are being upgraded almost daily and are therefore subject to change. Use the values for estimation *only*. If you have a critical calculation or determination to make, the value listed in this text may be correct. *But*, be sure to find the best known value and verify the one from this text.

James K. Hallenburg
Tulsa, Oklahoma

The Author

James K. Hallenburg is currently a retired geophysicist/petrophysicist living with his wife, Jaquelyn, in Tulsa, Oklahoma. Jim was born in Chicago in 1921, and has traveled to and worked in many corners of the world. During World War II, Jim was a bomber pilot and weather observer in the U.S. Army Air Corps. and still enjoys flying.

Following his stint in the military, Jim returned to his studies, graduating with a B.A. in physics from Northwestern University. Throughout his career, he has attended a number of schools to further his education and has taught courses at several of them. Jim spent 18 years as an engineer, designing geophysical systems for Schlumberger Well Services He was a Senior Engineer for the Mohole Project and Chief Engineer for the Western Company of North America. He also operated and owned Data Line Logging Company in Casper, Wyoming. Later, he was Manager of Applications Engineering for Century Geophysical in Tulsa. Jim has also been a consultant for the International Atomic Energy Commission.

Subsequently, Jim spent several years giving seminars in geophysics and formation evaluation. He is the author, co-author, and editor of several books and computer programs and has held office in numerous technical societies.

1

Introduction

1.1 Formation Evaluation

Formation evaluation is the process of using borehole and surface information in concert to evaluate the characteristics of subsurface formations. In doing so, the composition, character, and quality of the zone in which we are working, can be determined.

Formation evaluation is the process of:

1. Estimating recoverable reserves of hydrocarbon, gas, minerals, water, or any other formation material
2. Estimating the target material in place (which, of course, is different from the recoverable material)
3. Determining the lithology and geology of the target environment
4. Assessing the general geological environment (correlation, mapping, depths, thicknesses, identifying depositional features)
5. Detection of abnormal pressures (especially in petroleum exploration)
6. Evaluation of rock stress and other pre-mining parameters
7. Locating and quantifying fluid contacts
8. Fracture detection, porosity type, and amount determination
9. Learning the history of the formation

Formation evaluation for petroleum follows some predictable steps. The process for petroleum is aimed at determining, primarily, the water saturation of the target formation zone. We will learn the other information, such as rock type, as a byproduct of the saturation determination or as a deliberate separate effort. But, the determination of the saturations is the first concern. To do this, specific information will be obtained and examined, and the saturations then calculated. There are many ways to do this. The process is shown in Figure 1.1.

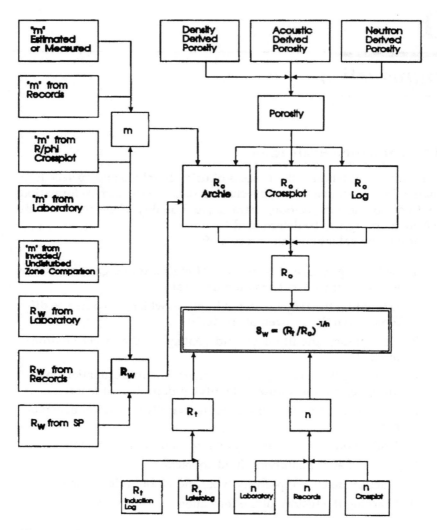

FIGURE 1.1
General procedure for evaluating hydrocarbon deposits.

1.2   Operation

In order to do these things, we must first see how the systems work and determine what their limitations and advantages are. In order to effectively predict what a system will do — what it will and will not see — we must first understand how it works.

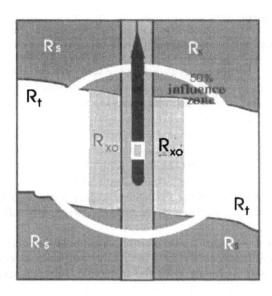

FIGURE 1.2
The borehole resistivity measurement situation.

We will examine the methods of obtaining this information, We will determine what, where, and to what degree the target is being measured. We will examine common methods for correcting data to eliminate unwanted information for each system. We will see why a system is preferred and what the shortcomings are. We will investigate the methods for reducing and interpreting the data, determine a general model, and determine specific things. We will see how we can adapt methods to new and different uses.

Once the systems are understood, the raw data can be processed. It must be reduced — it is seldom in a form that can be used unprocessed. That is, the raw signal values must be corrected for various unwanted and distorting factors, such as the presence of the borehole and the departure of the measurement geometry from ideal. This is because each measured volume is finite and will include more than the sample we wish to examine. See Figure 1.2. The corrected signal must then be converted to a useable form. For example, it is difficult to determine much about geological parameters with millivolts. If millivolts are converted to salinity, however, then the information can be used.

Finally, one must decide what the particular value from the reduced data means, especially when used in correlation with other measured values. This last is formation evaluation. This, then will be the process we will study.

1.3   Methods of Gathering Information

In the process of formation evaluation, we will use drilling operation logs, geophysical wireline logs, core analysis, and production type tests. Surface, airborne, and remote sensing methods may also be used, especially as preliminary indicators.

Drill cuttings are a good source of information. Because the samples are small and incomplete, and because the depths are uncertain, cuttings are a supplemental method. Depths from which the cuttings came are sometimes difficult to determine accurately because of the time lag between the cutting and their arrival at the surface. Also, larger pieces may travel to the surface at a different rate than small pieces. Clays are mostly dispersed in the liquid.

Cuttings are the bits and pieces which have been ground and cut out of the formation material by the drill bit. They have been mechanically damaged and washed severely with the borehole fluid. Descriptions of the washed drill cuttings are usually logged (tabulated) as a function of the drilling depth and then correlated with the core samples and wireline logs. They can be quite valuable if they are properly handled. Among the things we can learn from the cuttings are some of the details of the formation lithology, grain size, redox condition, and hydrocarbon presence. Many other things may be determined with varying degrees of uncertainty.

Other peripheral logs are mud logs (Figure 1.3), drilling time logs (Figure 1.4), cuttings analysis, mud analysis, core gamma logs (Figure 1.5) and core analysis logs (Figure 1.6). These are not wireline logs, but are less expensive ways of gathering additional formation information since they seldom require additional rig time. They are also usually made on the surface, where the environment is less hostile than down hole and they can usually be made while other operations are in progress. The amount of information which can be obtained is sometimes rather limited. This does not mean, however, that the information is valueless. It should be gathered and used and become part of the record.

This group of logs is often neglected in the descriptions and explanations of the sources of information for a project. They are valuable and must be considered. Some, such as the core analysis and the wireline logs, supply basic and necessary information. Many give information not available from any other source (drilling time and mud analysis). Other are used for correlation and verification (cuttings analysis and core gamma logs). Some serve dual purposes (drilling time logs and mud analysis). More details about the uses of some of these peripheral logs and sources are covered in later chapters.

The equipment to gather these items of information are usually sophisticated and designed to operate in the field. A good example of this type of instrument is the wireline logging truck (Figure 1.7). This is a complete,

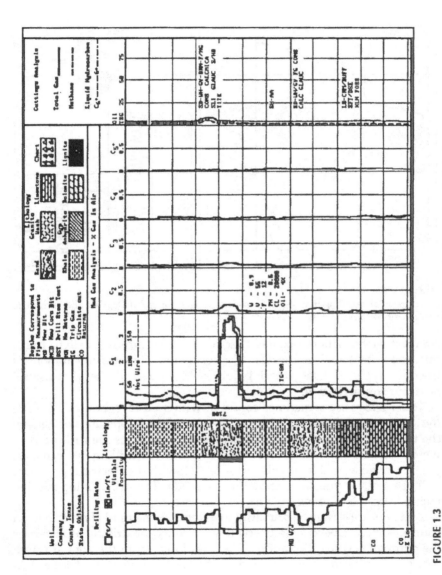

FIGURE 1.3
Typical mud log. (From Helander, D. D., *Oil, Gas, and Petrochem Equipment*, PennWell Books.

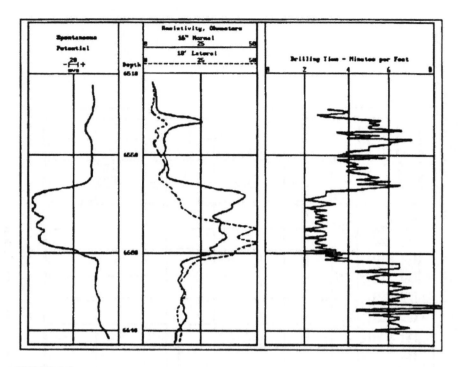

FIGURE 1.4
Comparison of a drilling time log with an electric log. Courtesy of Oil, Gas, and Petrochem Equipment, PennWell Books.

compact, mobile laboratory. It is designed to collect, process, deliver, transmit, and (often) analyze the desired information. The instrumentation for the other types of data-gathering devices are of a similar level of usefulness and sophistication.

Driller's logs are sometimes confused with mud logs, but they are not the same thing. The driller's log is a time log of the happenings on the rig: the trips, bit changes, mud changes, etc. This includes the drilling rate, which can be a valuable source of information. The drilling rate can alert you to possible trouble, it will correlate well with the electric logs, and it will often allow you to estimate missing data. The driller's log is often included on the mud log.

Mud logs are designed primarily to determine the mud character. Mud control is vital, especially in deep oil wells. Since the drilling mud carries the cuttings and the formation solids and fluids from the zone through which the drill is passing, it is an ongoing record of the subsurface events and some of the surface events. Much information may be gained about formation fluids and hydrocarbon presence. Mud logs are seldom used on mineral, water, and scientific projects.

Mud logs are time and depth records of the borehole fluid, the drilling mud. Typically, salinity, density, viscosity, additives, gas content, oil content,

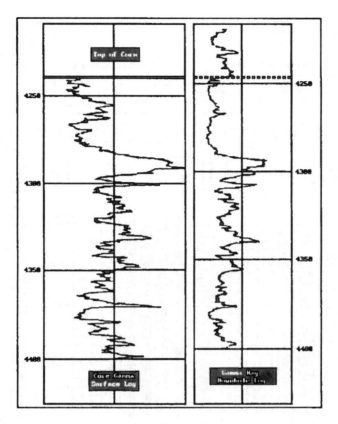

FIGURE 1.5
A surface core gamma log from Grayburg-San Andres, Andrews County, Texas. (From LeRoy, L. W. and LeRoy, D. O., *Core Analysis*, Core Laboratories, 1982. With permission.)

resistivity, and water loss can be logged, usually on the surface. These logs can furnish the analyst valuable information about the mud resistivity, temperature, depth of invasion, and general problems.

Wireline geophysical logs are the lowest cost per unit of information and the most dependable sources of high quality information for formation evaluation. Whether they are wireline logs, surface or airborne measurements, cross-hole (hole-to-hole) logs, or MWD logs, the modern trend is to offer more variety of physical, mechanical and chemical measurements and methods of interpreting them. At the same time, accuracy and reliability have increased and continue to increase markedly. These are used, among other things, to determine stratigraphy, lithology, porosity, and many other useful physical properties. Therefore, we will spend most of our effort learning the use of the wireline geophysical logging methods.

Wireline logs cost approximately ten times more than the drilling operation logs. They are, however, the least expensive source of detailed and precise data that we will be able to gather on a location. Thus, they are the

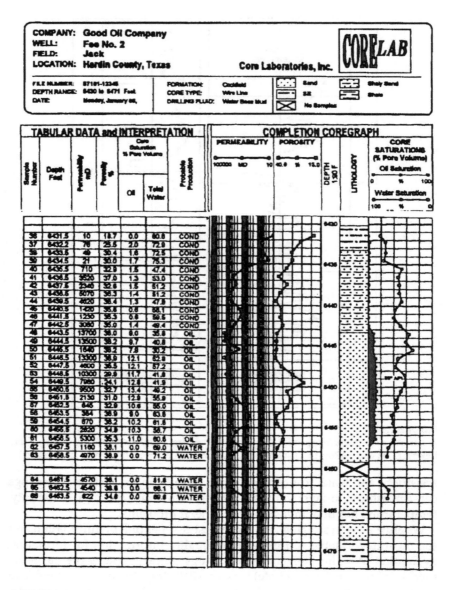

FIGURE 1.6
A typical core analysis log. (From LeRoy, L. W. and LeRoy, D. O., *Core Analysis*, Core Laboratories, 1982. With permission.)

primary method by which we determine the formation parameters. See Figure 1.8.

There are many types of geophysical measurement logs, mostly wireline logs, available from instrument suppliers and service contractors. These include electric logs, porosity logs, radioactivity logs, mechanical, and

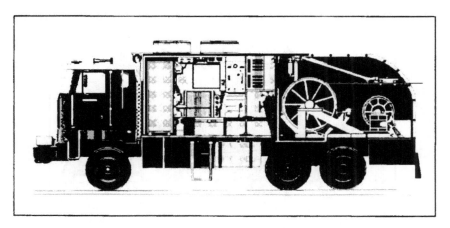

FIGURE 1.7
A typical well site mobile laboratory. (After Schlumberger Well Services, Inc., 1988.)

magnetic logs. These are available as airborne and surface methods, and downhole methods. These should always be used in combination and cross-correlated. The phenomenon that affects one will certainly affect the other, but in a different way. Each, however, will have severe limitations and distortions which must be accounted for.

Core information costs about ten to one hundred times as much as wireline logs for the information gathered. It is often the most detailed and exacting information we can get on the formation material. It is, however, very limited in scope and quantity. Usually, we cannot dispense with the need for good core analysis, especially in new or "wildcat" areas. Coring is covered in more detail in Chapter 6.

Cores and core analysis will be used to determine many of the parameters we cannot measure or have difficulty measuring with the other systems. Some of these things include formation permeability, the elements and details of lithology, a calibration of porosity in some cases, the exact fluid composition, the probable degree of hydrocarbon saturation, some of the formation mechanical and electrical properties.

Because of the high cost of cores and core analysis, the small sample size, and the long turnaround times, these methods are usually reserved for checking and verifying purposes.

Core analysis is one of the most valuable sources of accurate information. Its quality, however, depends upon the care with which the cores are taken, packaged and analyzed. They are no different, in this respect, than any other measurement method. There are as many examples of poor practice, mistakes, and carelessness with cores as with any other method of information gathering.

Sidewall coring is done with a wireline gun which fires hollow bullets into the wall of the hole. The bullets are then retrieved with the core inside. This method damages the core severely and the samples are small. It is a

FIGURE 1.8
A schematic of a petroleum logging setup.

second-choice method to use if the cores and/or the logs cannot supply the information. They are used routinely in the Texas Gulf Coast to look for hydrocarbon traces.

Production tests cost at least ten times as much as core analysis for the amount of information gathered. Therefore, they will be used much less frequently. As with the core analysis, some of the information we need can only be obtained from production test methods.

Geophysical measurements which are made with MWD are hole deviation, gamma ray, and resistivity. There is a rapidly growing list of others. Of these, the deviation measurements are among the most important.

Measurement while drilling (MWD) systems combine some of the advantages of drilling time logs with those of geophysical logs. Various production and geophysical measurements are made during the drilling process. These systems make geophysical and operational measurements during drilling. The instrument package is usually immediately above the drill bit. These systems originally were primarily methods to monitor the drilling process. They do, however, measure physical parameters.

The great advantage to MWD is that measurements are made and transmitted to the surface, or stored, as the drill progresses. All other methods require the drill and drill pipe to be brought to or near to the surface before measurements can be made. Thus, with MWD, vital information about the drilling process, such as hole deviation and drilling machinery monitoring,

TABLE 1.1

Average Chemical Composition of Some Formation Materials

	SiO$_2$	TiO$_2$	Al$_2$O$_3$	Fe$_2$O$_3$	FeO	MnO	MgO	CaO	Na$_2$O	K$_2$O	H$_2$O	P$_2$O$_5$	BaO	CO$_2$	SrO	C	SO$_3$	S	ZrO$_2$
Sandstone	78.33	0.25	4.77	1.07	0.30	—	1.16	5.50	0.45	1.31	1.63	0.08	0.05	5.03	—	—	0.07	—	—
Limestone	5.19	0.06	0.81	0.54	—	0.05	7.89	42.6	0.05	0.33	0.77	0.04	—	41.5	—	—	0.05	0.09	—
Dolomite	5.60	0.06	1.07	0.16	0.31	0.04	20.1	28.1	0.21	0.06	0.46	0.46	—	43.3	—	—	—	0.09	—
Gypsum	0.40	—	2.97	0.77	—	—	1.53	30.8	—	—	17.5	—	—	2.80	—	—	43.8	—	—
Norite	52.0	0.21	17.2	0.65	6.65	0.13	8.98	11.4	1.83	0.40	0.63	0.12	—	—	—	—	—	—	—
Synite	60.3	0.32	18.3	2.31	3.35	0.07	0.75	2.62	5.14	5.46	0.95	—	—	0.38	—	—	—	—	—
Granodiorite	73.4	0.2	14.1	0.7	1.7	0.02	0.4	2.1	3.4	3.5	0.3	—	—	—	—	—	—	—	—
Basalt	48.0	3.93	17.8	3.96	8.22	0.13	7.27	8.18	3.64	1.34	—	0.51	—	—	—	—	—	—	—
Gabro	51.2	1.59	14.5	2.91	11.7	0.14	4.33	7.65	3.21	1.05	0.56	0.23	0.01	0.01	0.01	—	—	0.06	—
Diorite	63.0	0.37	16.2	1.88	4.21	0.08	2.61	4.33	3.58	1.99	1.52	0.34	0.04	—	—	—	—	—	—
Dunite	40.1	0.28	0.57	1.77	8.40	0.10	45.5	0.23	0.03	0.05	2.12	—	—	—	—	—	—	—	—
Diabase	52.5	0.83	14.6	1.45	9.18	0.13	8.00	9.76	1.78	0.90	0.96	0.08	—	0.03	—	—	—	—	—
Granite	71.9	0.38	15.1	0.95	1.09	0.11	0.69	1.22	2.60	3.94	0.13	0.18	0.08	—	—	—	—	—	0.01
Glauconite	53.6	—	9.56	21.5	1.58	—	2.87	1.39	0.42	3.49	5.96	—	—	—	—	—	—	—	—
Kaolinite	45.6	0.80	37.4	0.69	0.03	—	0.34	0.48	0.20	0.63	13.8	—	—	—	—	—	—	—	—
Montmorillonite	52.4	0.12	18.3	2.58	0.21	—	3.93	1.87	0.66	0.31	19.8	—	—	—	—	—	—	—	—
Illite	50.4	0.33	27.0	3.57	1.07	—	2.61	0.30	0.40	6.48	7.68	—	—	—	—	—	—	—	—

is gathered within seconds or minutes after occurrence, rather than hours, days, or weeks after.

Data are transmitted to the surface via digitally encoded pulses with high information densities. Earlier, analog methods were used, and the information density was so low that most of the measurements were not feasible. Data are transmitted by pulse pressure changes in the mud column or by electrical transmission through electrical conductors in special drill pipe. The mud pulse transmission method was invented about 1950, by Arps who was a consultant in Oklahoma.

The disadvantages to MWD are its (sometimes) low information rate and the hostile environment. These situations, of course, are changing rapidly. MWD is feasible because of modern solid state electronic technology. Even ten years ago, much of the present capability was not possible.

Another method which is used primarily for engineering purposes is the penetrometer method. In this, a cone penetrometer and a friction sleeve are pushed into the earth by a truck-borne hydraulic system. The mechanical resistance to the cone and the friction on the sleeve are measured. Other physical parameters may also be measured, at the same time.

1.4   The Borehole Environment

The borehole environment is hostile. Measurements, observations, and examinations are made from a small diameter hole or a small quantity of material taken from that hole. The hole may be in poor condition; it may not be straight and it may be filled with a fluid in which we are probably not interested. In fact, the fluid will actually interfere with, dilute, or distort measurements. The ambient temperatures and pressures are usually high (they can be as high as or higher than 260°C and 2000 atmospheres). Physical conditions change drastically from the bottom of the hole to the surface.

1.5   Data Reduction

The data gathered by the various logging methods are in the parameters of the measurement. These may be counts per seconds, millivolts, microseconds, etc. For the end user, these units are relatively meaningless. The data measurements are usually reduced to values which have some meaning in the context of the information desired. Therefore, counts per second can be related, quantitatively, to percent uranium, porosity, or other units for which the log was designed. The same reduction process must be followed for every log.

The fundamental equations for data reduction of the logged information may have already been used (digitally or in an analog process) during logging, before the log is printed out. If the data have not been reduced, then the following equations are some of the necessary ones. If the data have been reduced, then one *must* determine how much has been reduced and which methods have been used, in order to properly evaluate the information.

The value of the spontaneous potential (SP) curve is related to the differences in salinities of the formation water and the mud filtrate:

$$E_{ssp} = K_{sp} \log \frac{R_{mf}}{R_w} \qquad (1.1)$$

where

E_{ssp} = the value of the SP curve (corrected with respect to the shale content of the zone)

K_{sp} = the SP conversion constant (in millivolts) and is temperature sensitive and always negative

R_{mf} = the electrical resistivity of the liquid portion of the drilling mud

R_w = the resistivity of the formation water

The reading of any resistivity curve, R_a, includes several components:

$$R_a = f(R_m, d_h, R_i, d_i, R_s, h, R_t) \qquad (1.2)$$

where

R_a = the apparent resistivity, as read by the measuring system

R_m = the drilling mud resistivity

d_h = the drill hole diameter

R_i = the electrical resistivity of the invaded zone

d_i = the diameter of the invaded zone

R_s = the electrical resistivity of the surrounding beds

h = the bed thickness

R_t = the true formation resistivity

The principle illustrated for resistivity by Equation 1.2, actually holds for any and all of the downhole measuring systems.

The bulk density, ρ_b, in a sedimentary zone depends upon the rock matrix density, ρ_{ma}, the amount of pore space (as "seen" by the density system), ϕ_p, and the density of the fluid filling the pore space, ρ_f:

$$\rho_b = \rho - \rho_{ma}\left(1 - \phi_p\right) + \rho_f \tag{1.3}$$

The neutron porosity, ϕ_N, and the neutron counting rate, N, are related semi-logarithmically and are affected by the borehole fluid and the tool constants, as well as the hydrogen content of the formation material.

The relationship for the acoustic systems resembles that for the density systems:

$$t = t_{ma}\left(1 - \phi_a\right) + t_f\phi_a \tag{1.4a}$$

$$\phi = \frac{\phi_a}{C_p} \tag{1.4b}$$

where

t	= the logged acoustic pressure wave travel time
t_f	= the travel time in the formation fluid
t_{ma}	= the travel time in the rock matrix material
ϕ_a	= the apparent (logged) porosity
C_p	= the compressional factor

The compressional factor is an arbitrary factor, derived from the shale reading, intended to approximately correct for log depth effects.

1.6   Reserve Calculations

Reserve calculations are used in all phases which eventually produce some portion of the formation. The basic relationship which is used in petroleum work is:

$$N = \frac{K_{res}hA\phi\left(1 - S_w\right)}{B_i} \tag{1.5a}$$

where

N = the initial oil or gas in place

K_{res} = a constant (7758 barrels per acre foot if oil, or 43560 STP$_i$ cubic feet per acre foot, if gas)

h = the zone thickness in feet, A is the drainage area in acres

ϕ = the fractional porosity

S_w = the water saturation

B_i = the initial formation volume factor

Note that the values of h, ϕ, and S_w can be determined with wireline logs.

$$N_P = NE_r \qquad (1.5b)$$

where N_p is the producible oil and E_r is the overall recovery efficiency.

In mineral work, the reserve calculations are formalized only for the gamma ray methods. Other methods are quite possible and many are feasible. The use of them, however, is not widespread enough that formal, standard relationships have been adopted. The reserve calculation for the gamma ray method is:

$$GT = KFW \Sigma(N) \qquad (1.6)$$

where G is the percent grade, T is the thickness of the mineral zone, K is the counting rate conversion factor, N is the corrected, normalized gamma ray counting rate at intervals through the mineral anomaly on the gamma ray curve, W is the width of the interval at which counting rates were determined (NW is the area of each rectangle used to integrate the curve; since W is constant, $W\Sigma(N)$ is approximately the area under the curve), and F is the product of all of the additional correction factors (hole size, mud attenuation, casing thickness, etc.). This relationship is suitable for uranium, thorium, and potassium.

The common shale-free equation used in petroleum work to determine the degree of water saturation of the pore space was suggested by Gerald Archie:

$$S_w^{-n} = \frac{R_t}{R_o} = \frac{R_t}{F_r R_w} \qquad (1.7)$$

where S_w is the water saturation as a fraction of the pore space, n is the saturation exponent and is near 2.0 in value, R_t is the true formation electrical resistivity, R_o is the formation resistivity when 100% water saturated, F_r is the formation resistivity factor, and R_w is the formation water resistivity.

TABLE 1.2

Approximate Values of the
Cementation Exponent, "*m*"

m	Porous Rock Type
1.3	Unconsolidated sandstone
1.4–1.5	Very slightly cemented sandstone
1.5–1.7	Slightly cemented sandstone
1.7–1.9	Moderately cemented sandstone
1.9–2.2	Highly cemented sandstone
2.2–	Carbonates

The value of the formation resistivity factor is related to the amount of pore space and its shape:

$$F_r = \phi^{-m} = \frac{R_t}{R_o} \tag{1.8}$$

where ϕ is the fractional porosity and *m* is the cementation exponent.

These relationships will be covered in detail in later chapters.

2

History

2.1 General

Geophysical investigation is as old as Man. We have always been interested in the world around us, assigning both metaphysical and prosaic properties to it. The examination of the mineral aspects of the earth certainly were in full swing by the time we were finding suitable stone to make axes, spear heads and knives. Evidence from prehistory points to an extensive knowledge of stone, use of oil seeps, metallic ore, and many other uses of early geophysics and geology. Prehistoric mines still exist in Europe, and those in North America date to before Europeans arrived to this continent. Early mining literature still exists. These early mining experiences have contributed to our language and customs. Indians of the West Coast of North America used petroleum seepage to waterproof baskets, in order to carry water.

The illustrations in this chapter are from literature furnished by Baroid Drilling Fluids, Inc. They have been included here to illustrate significant milestones in the history of drilling for oil, water, and minerals.

In the application of rotary drilling and drilling muds, the Spindletop field near Beaumont, Texas was among the first to use this combination commercially, in 1909 (NL Baroid, 1979). Prior to that, cable tool systems had primarily been used. Some cable tool rigs are still in use. Commercial muds were introduced in 1928. The rotary drilling techniques and the drilling muds have become much more sophisticated than those used in these historically important events.

The Chinese introduced drilling and the use of drilling muds in about 256 B.C. for the production of brine. Their methods changed little in the next 1900 years. It is highly probable that similar drilling methods were used in other places in the world, through history (Figure 2.1).

Geophysical and geological exploration and formation evaluation remained largely an empirical art until the nineteenth century. Certainly, the need for good methods was recognized. Witching for water and metals sometimes apparently worked and sometimes not. In 1869, Lord Kelvin logged a water well with a platinum resistance thermometer circuit (Allaud et al., 1977). This appears to be the start of the modern science of geophysics. Late in the nineteenth century, the voltages caused by redox currents surrounding metallic bodies at the water table were mapped.

FIGURE 2.1
An early Chinese drilling rig. (Courtesy of Baroid Drilling Fluids, Inc.)

Some ore deposits appear to have been found in this manner. The Schlumberger brothers began investigating electrical sounding methods in 1910. Anne Grüner Schlumberger (Schlumberger, 1982) speaks of her father, Conrad Schlumberger, who was Professor of Physics at the Ecole des Mines, experimenting with electrical methods in the basement laboratory in 1911. Before 1920, the Schlumbergers offered a service in which they detected the induced currents caused by an alternating magnetic field on the face of a mine drift. These currents were interpreted as a function of the resistivity of the formation material ahead of the face. A torsion balance was invented in 1890 and used by Baron Roland von Eötvos, Professor of Physics at the University of Budapest to detect salt domes in Hungary (DeGolyer, 1935). The Spindletop salt dome was discovered by means of a gravity survey in 1924. In 1923, a German crew did commercial refraction seismic surveys for the Mexican Eagle Oil Company and the Marland Oil Company. Use of the method had been conceived by L.P. Garret and Robert Welch of Houston in 1905. In 1848 Robert Mallet delivered a paper to the Royal Irish Academy entitled "On the Dynamics of Earthquakes: Being an Attempt to Reduce Their Observed Phenomena to the Known Laws of Wave Motion in Solids and Fluids." Mallet also used gunpowder explosions to generate noise for his seismicity experiments. In 1921, Marcel and Conrad Schlumberger logged, *in situ*, the resistivity of a coal bed from a

FIGURE 2.2
Springpole rigs used in Belgium in 1828 and in Pennsylvania. (Courtesy of Baroid Drilling Fluids, Inc.)

borehole in the Beseges Basin near Molieressur-Ceze in France (Allaud et al., 1977). The hole was 2500 feet (760 meters) deep and was cased to 1500 feet (460 meters).

At the time that the Schlumberger brothers began their borehole logging, the accepted theory was that the overburden pressure in sediments would collapse any pore space in the rock and the pore volume and permeability would be zero. Thus, the resistivity would thus be "infinite". These tests proved conclusively that the prevailing theory was incorrect.

In 1927 Conrad Schlumberger outlined the principles of "Electrical Coring" in a note entitled *Reserches Electriques Dans Les Sondages*. Henri Doll was assigned the responsibility for equipment design and testing. The first operation was at Diefenbach in Well no.2905, Rig no.7, September 1927.

The history of coring and cuttings collection and analysis goes back to the early Chinese (NL Baroid, 1979 and see Figure 2.1). Drilling method remained primitive into the 19th century (Figure 2.2), when "cable-tool" rigs were introduced (Figure 2.3). Rotary rigs were adopted in the early 20th Century (Figure 2.4). These early drillers apparently studied the cutting the drilling of their water wells to determine the quality of the zones. Early Chinese woodcuts show cable tool drilling rigs and sample collection. Downhole gamma ray measurements were made commercially in 1938. They were already an accepted surface method. Airborne gamma ray spectroscopy led the spectrographic field for many years. This was because of the bulk of the detectors and the analyzers. Neutron methods were introduced about 1950. As far as the author is aware, there is no surface

FIGURE 2.3
A cable-tool rig in Pennsylvania in the Drake period. (Courtesy of Baroid Drilling Fluids, Inc.)

method comparable to the neutron porosity. Pulsed neutron methods have been used in the laboratory and have been a valuable forensic tool for thirty years. Airborne and down hole gravity measurements became possible with the development, about forty years ago, of portable gravimeters.

For many years the studies of geology and geophysics were essentially separate. Geologists did not study geophysics and most geophysicists were recruited from the ranks of physicists and engineers. Geophysicists, likewise, did not associate with "loggers". We are beginning to accept now, that "logging" is a geophysical method, that geophysicists must know geology and that geophysics is an invaluable tool in geology. Therefore, we will attempt to combine these fields in this book and present a well-rounded picture of the result, formation evaluation.

Because of the importance and economics of geophysical logging methods, the subsequent history of formation evaluation is essentially that of the logging industry. This is not to belittle the vitally necessary surface, core, and cuttings methods. The fact is, geophysical logging is very cost effective.

Even though the spontaneous potential (the measurement of natural potentials in the earth) was well known and had been used on the surface since before the turn of the century, it was added to the downhole resistivity instruments only in 1931.

FIGURE 2.4
An early rotary rig at the time of the Spindletop boom. (Courtesy of Baroid Drilling Fluids, Inc.)

In the late 1960s, the mineral industry began to become interested in geophysical logging methods. This coincided with the upswing of the uranium exploration. Some geophysical methods were in regular use, but were not used extensively. The use of the gamma ray measurement for exploration and evaluation of uranium deposits *in situ* was quickly recognized as a tremendous saver of time and money.

Early gamma ray measurements were adopted directly from the petroleum industry (Figure 2.5). The mineral industry, however, had a long history of quantitative use of several of the methods which were used qualitatively in petroleum work. The gamma ray spectrograph had been used quantitatively in airborne and surface measurements for some time. Therefore, it was natural that the downhole gamma ray methods be used in a similar fashion. Thus, the tight calibration of the gamma ray measurement was a goal from the start, in uranium exploration. High sensitivity, accurately calibrated, stable, wide sensitivity range, digital systems were pioneered and perfected by the sedimentary mineral business. They were in common field use before 1976. Hard-rock mining became interested in logging methods during the early years of the 1980s. Scientific and environmental logging came into use during the late 1970s. Digital data systems,

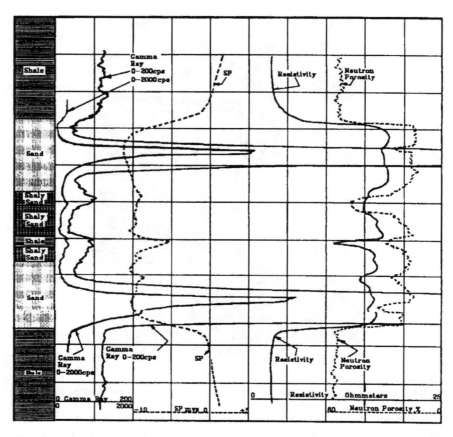

FIGURE 2.5
A digital mineral log through the oxidized zone of a uranium roll front in sediments,
Wyoming. (Courtesy Century Geophysical Corporation.)

which offer many advantages over analog systems, were introduced into
field use by the mineral exploration companies. These, of course, had to
wait for the practical, small computer and were introduced into commercial use in 1976, by Century Geophysical Corporation. Schlumberger introduced the oilfield version of these systems in 1981. The digital logging
systems have now almost completely supplanted the analog systems. Surface geophysical methods have been used for many years in both petroleum and mineral exploration (Figure 2.6). Recently, the effort has been to
coordinate and combine these with downhole methods.

For many years the studies of geology and geophysics were essentially
separate. We are now beginning to accept, that these two fields are different aspects of the same field, and both are necessary for satisfactory study
of the earth and formation evaluation. Because of the importance and economics of geophysical logging methods, the subsequent history of formation evaluation depends heavily on borehole geophysical logging. This is

FIGURE 2.6
An isopac map generated from mineral logs. (After Wyoming Geological Association.)

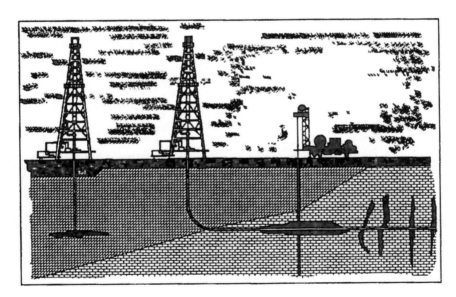

FIGURE 2.7
On-shore drilling technology in the 1990s.

not to belittle the value of surface, core, and cuttings methods. The fact is that logging is very cost effective.

Recent developments cover many areas and devices (Figure 2.7). The most significant, however, seems to be the adoption of digital techniques, on-site computers, and the great improvements in reliability, flexibility, overall quality, and presentation of the finished product. When this is coupled with modern communications, the result is a product which is complete, readable, and on-hand when and where it is needed.

3

Formation Material Characteristics

3.1 Petrophysics

Petrophysics is the science of determining the characteristics of rocks. With it one can identify reservoir rocks, identify and measure porosity systems, determine the degree of water and hydrocarbon saturation, estimate permeability, determine capillary pressure, and find and make use of measurements of formation materials physical properties, such as the electrical, nuclear, and mechanical properties. Rock materials may be identified *in situ* by their properties. Table 3.1 (see end of this chapter) lists the proportions of some of the components of a few common formation materials.

Almost all petroleum production is from sediments. There are few viable petroleum deposits which are not sedimentary. On the other hand, mineral deposits inhabit all formation types. Engineering, scientific, and environmental investigations involve all formation types. Hard-rock environments will be discussed. Probably 95% of the research and engineering in logging and other geophysical methods has concerned sedimentary formation materials, therefore, most of our examples will be from the sediments. Be aware, however, that the methods and explanations will apply, as well, to other rock types and other disciplines. Rock properties are important to petroleum, mineral, and engineering determinations. They are particularly important to formation evaluation and we will discuss the characteristics of the zones and materials in which we will be working. The materials which will be discussed are the rock materials in which ores, water, hydrocarbons, and other important minerals are found and the fluids which fill the voids in those materials. These properties are also important for engineering, scientific, and environmental purposes.

The interesting rock materials in the sediments are mostly clastic. These are gravel, sand (quartzose, arkosic), silt, mud, clay, or any mixtures of these. The consolidated equivalents of the clastics are sandstones, shales, and conglomerates. Figure 3.1 shows a typical arkose structure.

The rock materials may also be carbonate (illustrated in Figure 3.2. These may be limestone, dolomite, or mixtures and are usually organic in ultimate origin and often show evidence of their sources. They may also be any mixture of materials, organic and inorganic in origin, as shown in Figure 3.3. Occasionally sedimentary zones will contain massive evaporite materials, such as halite, gypsum, and potash.

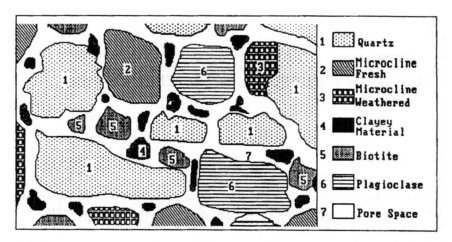

FIGURE 3.1
Arkose-type reservoir rock.

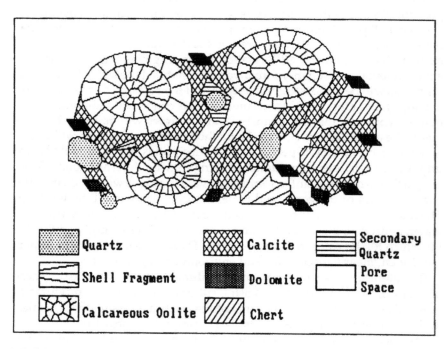

FIGURE 3.2
Porosity development in a complex oolitic limestone.

The hard rock materials are quartz, granites, basalt, feldspars, and many others. These hard rock materials (metamorphic and igneous) are often massive and may be fractured and intruded. Their fractures often indicate the mechanism of the fracturing, for example, Figure 3.4 depicts fractures

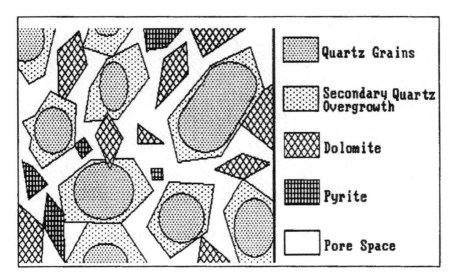

FIGURE 3.3
Dolomitic quartzite (Wilcox Sand, Oklahoma).

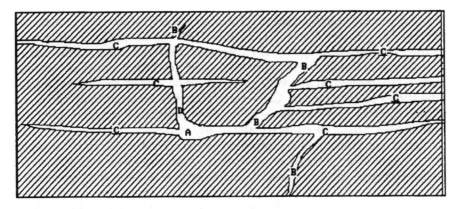

FIGURE 3.4
Carbonate rock, showing porosity derived from solution along joints and bedding planes;
A, vugs; B, joint channel; C, bedding plane channels; D, solution channel.

in a carbonate formed by solution. Figure 3.5 illustrates mechanical stress fracturing. Stress fracturing frequently precedes solution occurrences. Hard rock materials are occasionally found in petroleum investigation, but are largely incidental. They are found mostly in non-petroleum work. These porosity-forming mechanisms operate in the same way in massive carbonates and form the vessel for the hydrocarbons sought in this type rock. The sedimentary rock materials will be filled with fluids, which may be water, oil, gas, or a mixture of any or all of these fluids. *Pore space is the fraction of the void space or non-rock space compared to the bulk or total volume.*

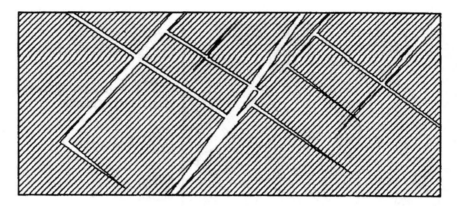

FIGURE 3.5
Carbonate rock showing porosity derived from fracturing and fissuring.

3.2   Rock Types — Sandstones

Sandstones are aggregations of lithified or unlithified mineral or rock particles of grain sizes 2 mm< >1/16 mm. They can be quartz, granite, gypsum, coral, shell, magnetite, glauconite, feldspar, mica, or hematite. Sandstone is usually sand grains cemented together; a siliceous composition often implied. They overlap the conglomerates on the coarse end and the shales on the fine end. Cementing materials are commonly silica, carbonates, clays, and iron oxides. Flagstone is sandstone which splits into thin layers. Freestone splits equally in any direction.

Sands and sandstones are important in petroleum and in mineral deposits because of their frequent occurrence (probably more than 80% of the near-surface formations material) and large pore volume and permeability. Porosity is normally 20 to 35% of total volume (depending upon sorting, packing, and cementing) and is usually interconnected (intergranular). It can be as low as 10% (or less) when cemented. Clays and shales will reduce the effective porosity to lower values.

Electrical resistivities in sandstones, sand, silt, and chert range from less than 0.1 ohmmeters in saline sands to 10^{14} ohmmeters in dry, quartz sand. Water-saturated sandstone electrical properties change with small and decreasing grain sizes. In the earth, resistivities will range from less than 1 ohmmeter up to about 1000 ohmmeters. See Table 3.3 (end of this chapter). The density of quartz is approximately 2.65 g/cc. The bulk density of dry "quartz" sand is from 1.68 to 2.41 g/cc. The matrix (rock) density of "quartz" sandstone is from 2.58 to 2.84 g/cc. The average is 2.544 g/cc. (Gearhart, 1982). See Table 3.4b at the end of this chapter. The potassium content of sandstones ranges from <1.1 to 5.1%. (See Table 3.5, at the end of this chapter) About 0.012% of all natural potassium is radioactive (^{40}K). It will emit a gamma ray of 1.5MeV energy upon disintegration. The average sandstone contains about 12% feldspar. Even orthoquartzite sands may contain 10% feldspar, particularly potash-bearing (orthoclase) varieties. A number of sandstones have been reported to contain an average of 50% feldspar. Acoustic travel time values are from 53.8 to about 100 microseconds/foot (176 to 328 microseconds per meter), average 57 microseconds per foot (187 µs/m) (Table 3.6 at end of this chapter). Graywacke sandstones are generally shaly and contain 33% or more easily destroyed minerals and rock fragments derived by rapid disintegration of basic igneous rocks, slates, and dark colored rocks. Such formations may contain a large quantity of potassium, but not in the quantity found in the arkosic sandstones (Gearhart, 1982).

Quartz has little or no natural radioactivity. Any radioactive emission in a quartz sandstone originates from the clay, feldspar, uranium, and thorium content. A typical gamma emission in quartzose sandstone is 25 APIg. The response of the neutron porosity tools is from 450 to 1000 APIn, depending upon the water and salt content. Arkosic sandstones are generally non-shaly and contain 25% or more feldspar, derived from the disintegration of acid igneous rocks. These sands have high potassium content and a higher radioactivity. Granitic sands are often quite high in radioactivity because of their potassium content and because granites are often the source rock for uranium compounds. Spontaneous potentials in sands and sandstones may be from zero to ±150 mvs), depending upon the salinity ratio of the formation water to the drilling mud and their shale content. Structure in sands may range from unconsolidated to highly cemented, granular, angular to rounded, clean to shaly, clean to cemented. They are composed of silica, carbonate, and other materials.

3.2.1 Carbonates

Carbonates are the calcareous limestone and/or dolomite. This group also includes siderite, magnesian limestones, dolomitic limestones, lithographic limestones, chalk, marble, travertine, crystalline limestone, and cherty limestone. Limestone is $CaCO_3$. In petroleum it is called limestone if it has more calcium than magnesium. Dolomite are $CaMg(CO_3)_2$ and mixtures with more magnesium than calcium are labeled dolomite. Limestones are usually initially deposited by organic action. They may be redeposited many times and in many different ways. They do also undergo other physical and chemical changes. Thus, they exist in many forms.

Carbonate structures may be firm, compact, and have high strength or loose and weak, depending upon their histories. They may occur in massive structures, pods in shales and sands, thin lamina, and they often contain sands, shales, and clays. Typical porosities in carbonates are from 0 to 15%, but may be 50% to 60% or more. Porosity can occur in many forms: inter-connected or isolated, intergranular (primary porosity), fragmented (pri-mary), inter-oolitic (primary), inter-vugular, (secondary), fractural (secondary), inter-crystalline, or any combination of these. Fractured and vugular porosity may easily consist of large voids, compared to the typical voids of the other types of porosity. The carbonate may be cemented and/or shaly and/or sandy.

At the time of deposit, carbonate sediments commonly have high poros-ities. By contrast, old carbonates contain only a few percent of pore space because of the processes of diagenesis. Carbonates will typically show rapid changes of porosity and thin porous horizons between low porosity layers. Along with the wide range of porosities is a wide diversity of pos-sible pore geometries. Choquette and Pray (1970) recognized 15 basic dif-ferent pore types. These included inter-particle, intra-particle, inter-crystalline, moldic, fenestral, fracture, and vugular types. Many carbon-ates will contain several types.

The uranium and thorium contents of carbonates normally are low. The gamma radiation typically will be from 5 to 15 APIg. The uranium content is usually lower in dolomite than in limestone, because the crystallization process tends to expel the uranium atoms from the crystal structure. Ura-nium deposits can occur in carbonate zones, however. Also, uranium com-pounds may deposit on the surface of carbonate boulders and on the surfaces of fractures, because of the high pH values of many carbonates. Spontaneous potentials in carbonates, zero to ± high (to ± 100 μs/m), depending upon the salinity ratio. These are often affected markedly by the typically high bulk resistivities in these rocks. Because of the frequent high resistivities of carbonate zones, the SP curve, in such zones, is typi-cally flat and featureless. The density of calcite is from 2.71 to 2.72 g/cc. That of limestones is 2.66 to 2.74 g/cc (2.69 g/cc). In dolomite, densities are found from 2.80 to 2.99 g/cc (2.85 g/cc). See Table 3.4a and b at the end of this chapter. Carbonate zone resistivities are usually high, 10 to about 6000 ohmmeters. Calcite may have 10^{12} to 10^{14} ohmmeters resistivity; limestone and dolomite, up to $>10^9$ ohmmeters. Of course, the water-filled porosity and/or the shale content can lower these drastically. See Table 3.3 at the end of this chapter. Acoustic travel times in calcite are usually 45.5 to 47.5 μs/ft (149 to 156 μs/m). In limestone, travel times are 47.7 to 53 μs/ft (156 to 174 μs/m), averaging 52 μs/ft (170 μs/m). In dolomite they are 40 to 45 μs/ft (131 to 148 μs/m), averaging 44 μs/ft (144 μs/m). Refer to Table 3.6 at the end of this chapter. Carbonate potassium contents run from 0 to 0.71%. See Table 3.5 at the end of this chapter. The thermal neutron capture cross section, sigma (Σ), for calcite, $\Sigma_{calcite}$ = 7.48 c.u., for limestone Σ_{ls} = 8.7 c.u. and for dolomite Σ_{dol} = 4.78 c.u. These values are listed in Tables 3.4a and b.

3.2.2 Clay Minerals

3.2.2.1 Clays

Clay minerals are fine-grained, layer-latticed minerals, most of which are hydrous alumino-silicates. They form in a weathering environment and reflect the type of parent mineral and the weathering conditions. They typically contain a large proportion of water and can grade into slurries, in some conditions. They have the property of electrically attaching metallic ions (cations).

The term "clay" denotes

1. **Size:** particles $<2\,\mu$ diameter,
2. **Ceramic:** plastic when wet and indurated when dry,
3. **Mineral:** lattice-layered silicates.

Other minerals may consist of clay-sized particles: quartz, feldspar, volcanic glass, oxides, organic matter, and some amorphous materials. Many of the physical properties of these approach those of clays, as their size decreases, because of the electrical properties (especially) of crystalline materials. Clay is an important component of water-based drilling mud. It (usually bentonite or red clay) is added to increase the fluid viscosity and to form mudcake to protect the permeable zones.

3.2.2.2 Shales

Shales are fine-grained terrigenous rocks (formed by the destruction of other rocks). They grade from clays on the small particle end to silts on the large particle end. Shales are generally considered to be mixtures of one or more clay minerals and silts, sands, and/or carbonates. They get many of their physical properties from their clay mineral. Modern sediments and many exposed shales of post Paleozoic age include one or more of the common clay minerals, such as illite, chlorite, kaolinite, and montmorillinite. Clay minerals of exposed Paleozoic shales and more deeply buried younger, shales consist only of illite and chlorite. These possibly represent the end, stable product of the other clay minerals when deeply buried for a long time.

The structure of clays to shale to slate will range from soft to hard. These materials are often laminar. They frequently contain fossil material and/or are carbonaceous. Clays and shales are characterized by their usually extremely low permeability. They very often constitute a permeability barrier. Layering and fissibility (property of splitting into layers) and the often laminated structures of shales reflect their frequent low energy depositional environment. Shale lithification is determined, to a large extent, by compaction. Time, weight, and temperature modify the shales, squeezing out the water and other (organic) fluids. The volume will normally change with compaction. If the liquids are trapped within the shale, the fluids, rather than the rock material, will eventually bear the overburden. This is called "over-pressuring" and can be detected by our formation evaluation methods.

Shales often contain potassium within their clay mineral and, commonly, electrically trap cations of uranium and thorium. Thus, they emit gamma radiation (3 to 10 times or more than quartzose sands). About 20% to 30% is from the ^{40}K and the rest from the uranium and thorium daughter elements. Gamma ray readings in shales normally range from about 75 to 150 APIg units (American Petroleum Institute gamma ray unit, an arbitrary unit). Shales can, however, be anomalously radioactive, either from migration of uranium into the shale or by fluid passage through fractures.

Thermal conductivities of clays and shales are lower than for sands and carbonates. They range from 2 to 4×10^{-3} (cal/cm sec) per (°C/cm). The actual value will depend largely upon the water content. See Table 3.7 at the end of this chapter. Resistivities are typically low in these materials because of their high water and ion contents. Shales may read up to 15 ohmmeters, but usually are much lower. Resistivities in siltstones, marls, and slates may be higher. Refer to Table 3.3 at the end of this chapter. Acoustic travel times are long in shales; $t = 60$ to 170 µs/ft (197 to 558 µs/m). Table 3.6 at the end of the chapter. Densities of shales will depend upon the water content. They generally run from 1.2 to 2.6 g/cc. (2.35 g/cc). Dry kaolinite has densities of 2.4 to 2.68 g/cc. Dry montmorillinite has 2 to 3 g/cc (2.63 g/cc average). See Table 3.4 at the end of the chapter. Thermal neutron capture cross section (Σ) of shales and clays will also depend upon the water content and upon the salinity. Some values in clays are kaolinite, dry, 13 Σ; montmorillinite, dry, 8 Σ. These are also listed in Table 3.4 at the end of this chapter.

The spontaneous potential (SP) of shale is zero by definition. Shale potential is defined as the zero of the SP differential measurement and all SP measurements are made in a positive or negative direction from this shale value.

3.2.3 Other Sedimentary Materials

Carbonaceous materials occur sometimes. These are coal (peat, lignite, bituminous, and anthracite), sapropel, black shale, oil shale, oil, and asphalt. These zones, in which we are directly interested for many mineral, engineering and for all petroleum work, will have some effective porosity and permeability. In mineral, engineering, and scientific work we are also interested in the hard-rock environments.

Extensive massive evaporite zones can exist in sediments, as well as evaporite materials dispersed within the other types of sediments. The properties of these materials are listed in the several tables of this chapter. Some of the evaporite materials are of commercial value, such as potash minerals, halite, and trona. Salt domes frequently have petroleum deposits associated with them. Salt domes and bedded salts have been of interest as possible repositories for petroleum, gases, high and low level radioactive wastes.

3.2.4   Hard-Rock Environments

Good information about formation evaluation and geophysical logging situations and examples in hard-rock environments is more difficult to find than that of sediments. This is because most of the research and engineering efforts have been expended by the petroleum industries. They are primarily interested in the sediments, since these are their source rocks.

Hard-rock environments usually are characterized by high resistivities, short acoustic travel times, moderate to high densities (hard rock types often show densities above 2.5 g/cc.), and low values of neutron capture cross section (Σ). The exceptions to these generalizations occur when the rock contains fractures. In that case, the water and alteration products in the fractures will alter the characteristics in much the same ways as in the sediments. They have radiation levels from virtually zero to about twice that of shales (disregarding any anomalous radiation).

Hard-rock environments normally show little or no useable SP. The SP is used, however, to detect water entry in geothermal wells. Granites, basalt, quartz, and other hard rocks are often massive, but may be fractured, altered, faulted, and intruded. They may have disseminated sulfides within them which, if concentrated enough, will drastically lower their electrical resistivity and other characteristics.

3.3   General Considerations

The various tables accompanying this chapter list the physical and some of the chemical properties of a number of formation materials. Please note that these parameters will change from time to time, as more data are gathered. Before using any of these values for exacting purposes, be sure to determine or confirm the value from an up-to-date source, such as the CRC *Handbook of Chemistry and Physics*.

Even though the chemical characteristics may not change for any given rock type, the mechanical arrangement of the sediment may change the physical values. This is primarily the reason for the range of values in the tables of natural material characteristics.

The degree of saturation of the sediment with liquids and, especially, with solutions of electrolytes will determine the bulk physical properties of the rock. This, of course, is what we take advantage of in geophysical logging for petroleum. Ionized electrolytes are electrically conductive.

Therefore, the degree of saturation of a sand with a solution of a salt in water will change the bulk conductivity of the sand. Displacing fluids, such as gases and petroleum liquids have extremely low electrical conductivities. Since these are electrically in parallel with the conductive solution, the net result is a change of bulk conductivity (and resistivity). The electrical resistivity, R_t, of the formation material is a function of the amount of fluid (especially the amount of water solution), the resistivity of the fluid, and the geometry of the fluid within the rock matrix. The type of rock material is of secondary importance to resistivity measurements.

This same type thing, to a different degree, must be considered with respect to the density, acoustic travel time, neutron capture cross section, and most of the other physical attributes we measure in bulk.

The grain size will also affect some of the physical parameters of the sediment. Broken crystals of highly resistive materials, such as quartz, will have unsatisfied electrical fields at their edges. Smaller grains have proportionally more surface area (more of these edges) than larger grains. Therefore, the electrical properties of these materials will change at small grain sizes. This can be seen on the logs, especially the electric logs. It can be effectively used to determine something of the history of the zone. Surface phenomena can also drastically affect the permeability of partially water-saturated, finer grain materials.

The surface area and the amount of pore space will depend upon the degree of packing. Maximum packing of the grains will result in more grains per unit area. This will increase the surface area per unit volume, the density, the acoustic velocity, and will decrease the porosity and neutron capture cross section. The sorting of the grain sizes will affect all of these properties, also. The more poorly sorted sands have lower porosities and more rock material per unit volume, because of the filling of the large grain pore spaces by smaller grains.

3.3.1 Compaction and Overpressure

Compaction by the overburden will also change some of the rock properties. Resistivity and density will increase, while travel time and neutron cross section will decrease because of the expulsion of water by the compaction process.

Compaction of the shales will force some of the contained water and other fluids out of the shale. These fluids will migrate to the surrounding sands. If the sands are hydraulically sealed by the shales, the excess fluids will begin to support the overburden, instead of the sand grains supporting it. This, too, will change all of the physical characteristics of the sand and, perhaps, the shale. They will then depart from the normal gradual change with depth. This is an overpressured situation and can be detected on the logs. Overpressure is further discussed in Chapter 4.

3.4 Fluids

3.4.1 Water Characteristics

Resistivity (the reciprocal of conductivity) of formation water is low to medium, depending upon the salinity. Its range is 0.02 to 50 ohmmeters. Gamma radiation is usually low (0 to 5 APIg). The primary exception is in potassium solutions. The density, depending upon the salt content and temperature, is 0.9 to 1.3 g/cc. Travel time also depends upon the salinity. It is 160 to 210 microseconds per foot (524 to 689 microseconds per meter). Its dielectric constant is about 80.

Water is the most common formation fluid. It is a very unusual zone which does not have some water. In general, the pore spaces will be filled with water; or the water saturation (saturation is the fraction of the pore space filled with the fluid) value, S_w is 100%. The water will have electrolytes in it, which will be dissociated into their component ions. These ions will move in response to an electric field. This is a flow of electric current. Energy must be supplied to do this work. The impedance to the flow is electrical resistance.

The measure of the ability of the ion to move is its mobility. Roughly, the larger and more massive an ion is, the lower is its mobility; the more energy it will require to move it. The more energy it has available in the form of heat, the less electrical energy it will require to move it.

If the temperature is not known, the resistivity measurement of a solution is meaningless. It is assumed that the temperature is formation temperature when formation resistivities and formation water resistivities are given. Thus, we will say that the electrical conductivity is higher at higher temperatures. See Table 3.2 and Figure 3.6. *It can be seen from this that it is necessary to know the temperature at which a solution resistivity measurement was made.*

It is important to know the temperature effect upon a solution because it will affect our electrical resistivity measurements downhole. Other measurements are also affected by the temperature: the SP, the density, and the neutron measurements, for example; but it is the electrical resistance measurement which shows the most important effects of salinity. We will start with that one.

We can usually easily measure the fluid (i.e., mud) resistivity at the surface, but, the chances are that it will not be at formation nor downhole temperature. Thus, the measurement will be in error. It is easy, however, to make a correction for the change of resistivity due to temperature differences. We also can get a sample of the formation water from the formation in the hole we are examining or from one nearby. A farmer may have a deep-water well or there may be a disposal well or an abandoned well nearby. We will undoubtedly have to make a temperature correction.

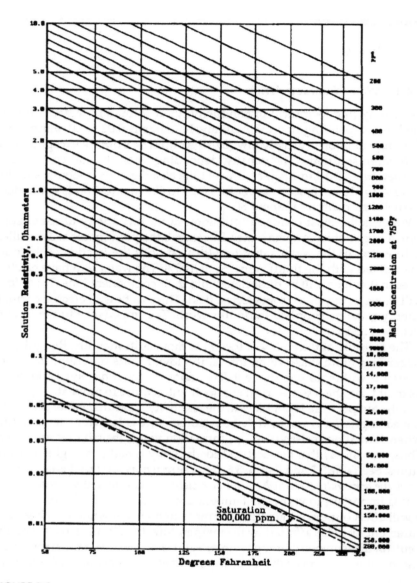

FIGURE 3.6
Resistivity, salinity, temperature characteristics of NaCl solutions. (After Schlumberger Well Services, Inc.)

The first step is to determine the downhole temperature. The most reliable way is to measure it. We will discuss temperature logging later but it is routine to measure the borehole temperature for resistivity logging purposes. We even have systems which will measure the formation temperature. However, they are expensive. For most purposes we can make an approximation which will be satisfactory. Anyway, we are usually concerned more with the borehole temperature than with the actual formation temperature.

The maximum temperature of the borehole can easily and cheaply be measured with a maximum reading thermometer in the cable head or with a circuit in the tool. We can assume that the maximum temperature is the bottom-hole temperature (BHT). The top of the hole is seldom at zero degrees. In a working well, we can take the temperature of the mud from the flow line or return line (T_{fl}). The difference between these two,

$$T_{diff} = BHT - T_{fl} \tag{3.1}$$

is the temperature change from top to bottom. If we divide this by the total depth of the well (TD),

$$\Delta T = \frac{BHT - T_{fl}}{TD} \tag{3.2}$$

we will have the average borehole temperature gradient, ΔT. The units will be "degrees F per foot" (°F/ft) or "degrees C per meter" (°C/m), depending upon the system in which we are working. See Figure 3.7.

Now we can calculate the probable borehole temperature at any level in the hole that we wish. This will be T_{form}:

$$T_{form} = \Delta TD_{form} + T_{fl} \tag{3.3}$$

The value of T_{fl} must be added because we are not starting the measurement at zero degrees at the surface. T_{form} is the probable borehole temperature at the level at which we are working. It is not the formation temperature. The formation temperature has been changed by drilling; it is, however, nearly the temperature at which we will measure formation resistivity.

Once we know the approximate temperature where we will work, it is necessary that all resistivities be corrected for that temperature. This may be done with the following approximation:

$$R_2 = \frac{R_1(T_{F1} + 6.77)}{T_{F2} + 6.77} \tag{3.4}$$

if the temperatures, T_{F1} and T_{F2}, are in degrees Fahrenheit or,

$$R_2 = \frac{R_1(T_{C1} + 21.3)}{T_{C2} + 21.3} \tag{3.5}$$

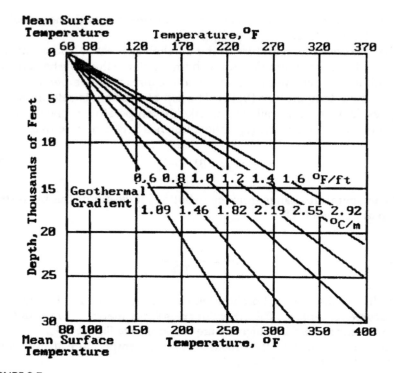

FIGURE 3.7
Estimation of temperature of the borehole at formation level. (After Schlumberger Well Services, Inc.)

if the temperatures, T_{C1} and T_{C2}, are in degrees Celsius. This approximation is good for the range in which we will be working. The value "7°F (or 22°C)" is sometimes used instead of 6.77°F (or 21.3°C). The difference is small and usually inconsequential. The value of "7°F (or 22°F)" may, therefore, of course, be used. **Always correct the values of R_m, R_{mc}, R_{mf}, and R_w for temperature effects before using.**

Formation temperature can be estimated from the borehole temperature measurement by making several measurements in the borehole at time intervals and extrapolating the trend to infinity. To do this, first make at least three temperature measurements in the borehole at intervals spaced in time (i.e., make a measurement on each of three consecutive runs). Plot these on the linear scale of a semi-logarithmic grid, against the value of

$$\frac{\Delta t}{t + \Delta t} \tag{3.6}$$

where t is the circulation time in hours. Δt is the time when the temperature measurement was made, in hours, after circulation was stopped. Extrapolate the trend to the value of:

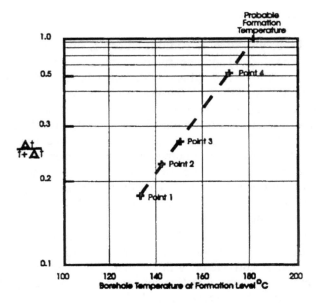

FIGURE 3.8
Estimation of formation temperature. (Courtesy of Western Atlas Logging Services Division of Western Atlas International, Inc.)

$$\Delta \frac{t}{t + \Delta t} = 1 \qquad (3.7)$$

The temperature value indicated on the linear scale, at the extrapolation, is the probable formation temperature. This is illustrated in Figure 3.8.

3.4.2   Ionic Conduction in Water

The conduction of electrical current in the pore fluid is due to the fact that electrolytes (salts, etc.) will dissociate upon solution in water. When an electric field, natural or artificial, is present, the ions will move in response. The positive ions (metallic ions; cations) will migrate in the negative direction (toward the negative electrode (cathode)). The negative ions (nonmetallic; anions) will migrate in the positive direction (toward the positive electrode (anode)).

Ionic conduction is also present in clays and shales; the most common conductive solid-formation materials. Metallic or electron conduction does not occur often enough to be a significant factor in this discussion. Electron conduction will occur in massive sulfides and in some high-rank coals.

Ionic conductance occurs because:

1. An electrolyte, such as sodium chloride, dissociates into posi-
    tively charged (+) and negatively charged (−) ions upon solution
    in a suitable solvent (the solvent must have a high dielectric
    constant). The only such solvent which we will consider, at this
    time, is water. The major solutes (electrolytes) in drilling mud
    and formation waters are $NaSO_4$, KCl, $CaCO_3$, $NaHCO_3$, $MgCl_2$,
    and $MgSO_4$. Note also, that metals immersed in a high dielectric
    solvent will undergo a somewhat similar process. A metal in
    water, for example, starts to go into solution as cations. If no
    process removes these ions, a cloud of cations will form in the
    solution, near the metal surface, finally saturating the solution
    and impeding further solution.

    Most of our work will be with eNaCl. That is, the character-
    istic that the electrolyte and solution would have if it were NaCl
    in water. A conversion will be made, if needed. We do not
    normally have the effective means, with logging systems, of
    determining the ion types downhole, within the formation. Core
    and fluid sampling analyses are employed to gain this informa-
    tion. Specific ion electrodes are available and are often quite
    good. Downhole, they measure in the borehole fluid, however.
    This is usually not very representative of the formation fluid.
    Measured solution resistivities are always given as eNaCl val-
    ues, unless specifically stated otherwise.

2. Electric fields will cause a migration of the cations to the negative
    direction and of the anions to the positive direction. This con-
    stitutes an electric current flow in the solution. The rate at which
    these ions migrate will depend upon their respective valences
    and masses. This aspect is termed the mobility of the ion.

3. The electric field may be set up by any one or more of several
    natural, induced, or interfering mechanisms. It may be due to
    an imbalance of the ion types in adjoining or the same zones. It
    may be due to an electrical circuit and electrodes deliberately
    placed in the formation. It may also be due to stray natural or
    artificial fields.

The electrical resistivity of a given solution will depend upon the concentration of the ions (activity) of the salts in solution and their types and
temperature (mobilities). The resistivity of any solution of salts may be
determined from the eNaCl value if their ion types and concentrations are
known. This process uses "Dunlap multipliers" (Schlumberger, 1986). See
Figure 3.9. Enter the sum of the concentrations of the ion types on the horizontal axis. Move upward, along the constant concentration line to a line
denoting the trend of an ion type. Move horizontally to the left axis and

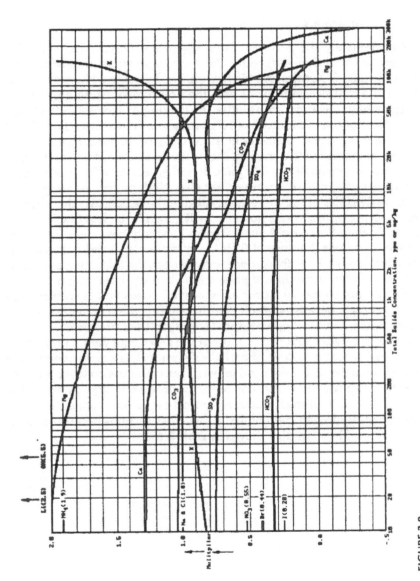

FIGURE 3.9
Equivalent NaCl concentrations from ionic concentrations; Dunlap multipliers. (After Schlumberger Well Services, Inc.)

determine the multiplier for the equivalent NaCl concentration of that ion. Multiply the concentration of each ion type by its appropriate multiplier. The sum of these equivalent concentrations will be the eNaCl concentration.

3.4.3 Hydrocarbons and Gases

Hydrocarbons are the petroleum oils and gases. Other gases will also occur in the formation reservoirs. These may be carbon monoxide, CO, carbon dioxide, CO_2, nitrogen, N_2, helium, He, hydrogen, H_2, and others.

All of the hydrocarbon fluids and the non-hydrocarbon gases have extremely high electrical resistivities. At logging or formation temperatures, these may be of the order of 10^9 ohmmeters or more. See Table 3.3 at the end of this chapter. Densities of this group, for the most part, are low. Table 3.4, at the end of this chapter, shows the densities of some of the simpler hydrocarbon materials. Oils have densities near 0.8 g/cc. Their densities will depend upon their molecular size. High molecular weight oils may have densities of 1 or more g/cc. All gases have densities that depend upon their pressures. These, however, will generally be less than those of the oils. Even in their liquid phases, gases generally have densities lower than the liquid oils (because of their lower molecular weight). Hydrocarbons which are gases at the surface, may exist in the formation as liquids if the ambient temperature and pressure are in the proper ranges.

Fluids do not support acoustic shear waves. Gases will absorb enough acoustic energy, because of their compressibility, that conventional logging equipment will not operate satisfactorily in their presence.

None of the hydrocarbons nor most of the gases emit gamma radiation. Radon, which occurs as three, short half-life isotopes, is a gamma emitter. It can dissolve in the other fluids and impart a gamma emission to the mixture.

Neutron capture cross sections of the hydrocarbon liquids are of the same order as that of fresh water. Hydrocarbon gases, in spite their high proportional hydrogen content, have low cross sections because of their low densities. These are also shown in Table 3.4.

TABLE 3.1a

Physical Parameters of Some Representative Formation Materials

Name	Formula	ρ_{FDC} g/cc	ϕ_{snp} p.u.	ϕ_{cal} p.u.	t_c μs/ft	t_s μs/ft	P_e b/ele	U b/cc	ε f/m	t_p ns/m	GR API	Σ c.u.
Silicates												
Quartz	SiO_2	2.64	−1	−2	56.0	88.0	1.81	4.79	4.65	7.2	—	4.26
β-Cristobalite	SiO_2	2.15	−2	−3			1.81	3.89			—	3.52
Opal (3.5% water)	$SiO_2(H_2O)_{.1209}$	2.13	4	2	58.		1.75	3.72			—	5.03
Garnet	$Fe_3Al_2(SiO_4)_3$	4.31	3	7			11.09	47.80			—	44.91
Hornblende	$Ca_2NaMg_5Fe_4AlSi_8O_{22}(O,F)_2$	3.20	4	8	43.8	81.5	5.99	19.17			—	18.12
Tourmaline	$NaMg_3Al_6B_3Si_6O_2(OH)_4$	3.02	22	22			2.14	6.46			—	7449.82
Zircon	$ZrSiO_4$	4.50	−1	−3			69.01	311.			—	6.92
Carbonates												
Calcite	$CaCO_3$	2.71	0	−1	49.0	88.4	5.08	13.77	7.5	9.1	—	7.08
Dolomite	$CaCO_3MgCO_3$	2.88	2	1	44.0	72.	3.14	9.00	6.8	8.7	—	4.70
Ankerite	$Ca(Mg,Fe)(CO_3)_2$	2.86	0	1			9.32	26.65			—	22.18
Siderite	$FeCO_3$	3.89	5	12	47.		14.69	57.14	6.8–7.5	8.8–9.1	—	52.31
Oxides												
Hematite	Fe_2O_3	5.18	4	11	42.9	79.3	21.48	111.27			—	101.37
Magnetite	Fe_3O_4	5.08	3	9	73.		22.24	112.98			—	103.08
Geothite	$FeO(OH)$	4.34	50+	60+			19.02	82.55			—	85.37
Limonite	$FeO(OH)(H_2O)_{2.05}$	3.59	50+	60+	56.9	102.6	13.00	46.67	9.9–10.9	10.5–11.0	—	71.12
Gibbsite	$Al(OH)_3$	2.49	50+	60+			1.10				—	23.11
Phosphates												
Hydroxyapatite	$Ca_5(PO_4)_3OH$	3.17	5	8	42.		5.81	18.4			—	9.60
Chlorapatite	$Ca_5(PO_4)_3Cl$	3.18	−1	−1	42.		6.06	19.27			—	130.21
Fluorapatite	$Ca_5(PO_4)_3F$	3.21	−1	−2	42.		5.82	18.68			—	8.48
Carbonapatite	$(Ca_5(PO_4)_3)_2CO_3H_2O$	3.13	5	8			5.58	17.47			—	9.09

TABLE 3.1a (continued)

Physical Parameters of Some Representative Formation Materials

Name	Formula	ρ_{FDC} g/cc	ϕ_{mp} p.u.	ϕ_{cal} p.u.	t_c μs/ft	t_s μs/ft	P_e b/ele	U b/cc	ϵ f/m	t_p ns/m	GR API	Σ c.u.
Feldspars — Alkali												
Orthoclase	$KAlSi_3O_8$	2.52	-2	-3	69.		2.86	7.21	4.4-6.0	7.0-8.2	~220	15.51
Anorthoclase	$KAlSi_3O_8$	2.59	-2	-2			2.86	7.41	4.4-6.0	7.0-8.2	~220	15.91
Microcline	$KAlSi_3O_8$	2.53	-2	-3			2.86	7.24	4.4-6.0	7.0-8.2	~220	15.58
Feldspars — Plagioclase												
Albite	$NaAlSi_3O_8$	2.59	-1	-2	49.	85.	1.68	4.35	4.4-6.0	7.0-8.2	—	7.47
Anorthite	$CaAl_2Si_2O_8$	2.74	-1	-2	45.		3.13	8.58	4.4-6.0	7.0-8.2	—	7.24
Micas												
Muscovite	$KAl(Si_{3Al0}10)(OH)_2$	2.82	12	20	49.	149.	2.40	6.74	6.2-7.9	8.3-9.4	~270	16.85
Glauconite	$K_2(Mg,Fe)_2Al_4(Si_4O_{10})_3(OH)_2$	~2.54	~23	~38			6.37	16.24				24.79
Biotite	$K(Mg,Fe)_3AlSi_3O_{10}(OH,F)_2$	~2.99	~11	~21	50.8	224.	6.27	18.75	4.8-6.0	7.2-8.1	~275	29.83
Phlogopite	$KMg_3AlSi_3O_{10}(OH,F)_2$				50.	207.						33.3

Courtesy of Schlumberger Well Services, Inc.

TABLE 3.1b

Physical Parameters of Some Representative Formation Materials

Name	Formula	ρ_{FDC} g/cc	ϕ_{amp} p.u.	ϕ_{cal} p.u.	t_c μs/ft	t_o μs/ft	P_e b/elec	U b/cc	ε f/m	t_p ns/m	GR API	Σ c.u.
Clays												
Kaolinite	$Al_4S_4O_{10}(OH)_8$	2.41	34	37			1.83	4.44	~5.8	~8.0	80–130	14.12
Chlorite	$(Mg,FE,AL)_6(Si,Al)_4O_{10}(OH)_8$	2.76	37	52			6.30	17.38	~5.8	~8.0	180–250	24.87
Illite	$K_{1-1.5}Al_4(Si_{7-6.5}Al_{1-1.5})20\ O_{20}(OH)_4$	2.52	20	20			3.45	8.73	~5.8	~8.0	250–300	17.58
Montmorillinite	$(Ca,Na)_7(Al,Mg,Fe)_4(Si,Al)_8O_{20}(OH)^4(H_2)_n$	2.12	40	44			2.04	4.04	~5.8	~8.0	150–200	14.12
Evaporites												
Halite	$NaCl$	2.04	-2	-3	67.0	120	4.65	9.45	5.6–6.3	7.9–8.4	—	745.2
Anhydrite	$CaSO_4$	2.98	-1	-2	50		5.05	14.93	6.3	8.4	—	12.45
Gypsum	$CaSO_4(H_2O)_2$	2.35	50+	60+	52		3.99	9.37	4.1	6.8	—	18.5
Trona	$Na_2CO_3NaHCO_3H_2O$	2.08	24	35	65		0.71	1.48			—	15.92
Tachydrite	$CaCl_2MgCl_2)_2(H_2O)_{12}$	1.66	50+	60+	92		3.84	6.37			—	406.02
Sylvite	KCl	1.86	-2	-3			8.51	15.83	4.6–4.8	7.2–7.3	500+	564.57
Carnalite	$KClMgCl_2(H_2O)_6$	1.57	41	60+			4.09	6.42			~220	368.99
Lanbenite	$K_2SO_4(MgSO_4)_2$	2.82	-1	-2			3.56	10.04			~290	24.19
Polyhalite	$K_2SO_4MgSO_4(CaSO_4)_2(H_2O)_2$	2.79	14	25			4.32	12.05			~200	23.70
Kainite	$MgSO_4KCl(H_2O)_3$	2.12	40	60+			3.50	7.42			~245	195.14
Kieserite	$MgSO_4H_2O$	2.59	38	43			1.83	4.74			—	13.96
Epsomite	$MgSO_4(H_2O)_7$	1.71	50+	60+			1.15	1.97			—	21.48
Bischofite	$MgCl_2(H_2O)_6$	1.54	50+	60+	100		2.59	3.99			—	323.44
Barite	$BaSO_4$	4.09	-1	-2			266.82	1091.			—	6.77
Celestite	$SrSO_4$	3.79	-1	-1			55.19	209.			—	7.90

TABLE 3.1b (continued)

Physical Parameters of Some Representative Formation Materials

Name	Fomula	ρ_{FDC} g/cc	ϕ_{snp} p.u.	ϕ_{cal} p.u.	t_c μs/ft	t_s μs/ft	P_e b/elec	U b/cc	ε f/m	t_{pl} ns/m	GR API	Σ c.u.
Sulfides												
Pyrite	FeS_2	4.99	-2	-3	39.2	62.1	16.97	84.68			—	90.10
Marcasite	FeS_2	4.87	-2	-3			16.97	82.64			—	88.12
Pyrrhotite	Fe_7S_8	4.53	-2	-3			20.55	93.09			—	94.18
Sphalerite	ZnS	3.85	-3	-3			35.93	138.33	7.8–8.1	9.3–9.5	—	25.34
Chalcopyrite	$CuFeS_2$	4.07	-2	-3			26.72	108.75			—	102.13
Galena	PbS	6.39	-3	-3			1631.37	10424.			—	13.36
Sulfur	S	2.02	-2	-3	122		5.43	10.97			—	20.22
Coals												
Anthracite	$CH_{.358}N_{.009}O_{.022}$	1.47	37	38	105		0.16	0.23			0–5	8.65
Bituminous	$CH_{.793}N_{.015}O_{.078}$	1.24	50+	60+	120		0.17	0.21			0–5	14.30
Lignite	$CH_{.849}N_{.015}O_{.211}$	1.19	47	52	160		0.20	0.24			1–10	12.79

Courtesy Schlumberger Well Services, Inc.

TABLE 3.2

Ion Mobilities

Salt	Cation	Anion	Cations Anions Dilute Solutions		K_p	Cations Anions 0.1 M Solutions		K_p
NaCl	Na	Cl	43.5	65.5	−11.6	36.25	55.5	−12.7
KCl	K	Cl	64.6	65.5	−0.4	55.0	55.5	−0.1
CaCl$_2$	Ca	Cl	51.6	65.5	−19.6	31.5	55.5	−26.4
MgCl$_2$	Mg	Cl	45.0	65.5	−22.5	27.5	55.5	−29.1
Na$_2$SO$_4$	Na	1/2SO$_4$	43.5	67.9	5.0	36.25	40.0	12.3
K$_2$SO$_4$	K	1/2SO$_4$	64.5	67.9	13.4	55.0	40.0	21.3
CaSO$_4$	1/2Ca	1/2SO$_4$	51.6	67.9	−7.9	31.5	40.0	−6.9
MgSO$_4$	1/2Mg	1/2SO$_4$	45.0	67.9	−11.7	27.25	40.0	−10.7
NaHCO$_3$	Na	HCO$_3$	45.3	46.7	2.2	—	—	—
Na$_2$CO$_3$	Na	1/2CO$_3$	45.3	60.0	7.5	36.25	38.0	13.5
K$_2$CO$_3$	K	1/2CO$_3$	64.6	60.0	16.0	55.0	38.0	22.4
Ca(HCO$_3$)$_2$	1/2Ca	HCO$_3$	51.6	46.7	−12.3	—	—	—
CaCO$_3$	1/2Ca	1/2CO$_3$	51.6	60.0	−4.4	31.5	38.0	−5.4
Mg(HCO$_3$)$_2$	1/2Mg	HCO$_3$	45.0	46.7	−15.2	—	—	—
MgCO$_3$	1/2Mg	1/2CO$_3$	45.0	60.0	−8.3	27.5	38.0	9.6
H$_2$SO$_4$	H	SO$_4$	315	67.9	42.4	—	—	—
HCl	H	Cl	315	65.5	37.9	—	—	38.3
H$_2$S	H	HS	315	42.5	44.0	—	—	—
H$_2$CO$_3$	H	HCO$_3$	315	46.7	46.6	—	—	—
NaOH	Na	OH	43.5	174	34.7	—	—	—

Mobility units are: cm^2/ohm, units of K_p are millivolts.
Courtesy Colorado School of Mines. After Dakhnov, 1959.

TABLE 3.3

Electrical Resistivities of Minerals

Mineral	Resistivity Ohmmeters
Sulfur	10^9–10^{16}
Petroleum	10^9–10^{16}
Biotite	10^{14}–10^{15}
Sylvite	10^{14}–10^{15}
Mica	10^{14} or more
Halite	10^4–10^{14} or more
Quartz	10^{12}–10^{14}
Muscovite	10^{11}–10^{12}
Calcite	10^7–10^{12}
Cinnabar	10^6–10^{10}
Sheelite	10^6–10^{10}
Anhydrite	10^4–10^{10}
Limonite	10^6–10^8
Sphalerite	10^5–10^7
Bauxite	10^2–10^6
Hematite	2.1×10^{-3}–10^6
Coal, Bituminous	10–10^6
Basalt	8×10^2–10^5
Gabbro	8×10^2–10^5

TABLE 3.3 (continued)

Electrical Resistivities of Minerals

Mineral	Resistivity Ohmmeters
Diabase	$8 \times 10^2 - 10^4$
Gneiss	$10^2 - 10^4$
Coal, Sub-bituminous	$10^2 - 10^4$
Dolomite	$1 - 7 \times 10^3$
Limestone, dense	$80 - 6 \times 10^3$
Lignite	$<1 - 4 \times 10^3$
Conglomerates	$<3 \times 10^3$
Siderite	$10 - 10^3$
Sand	$<10^3$
Slate	$<10^3$
Marl	<600
Siltstone	<300
Peat	$<1 - 300$
Argillite	$1 - 300$
Braunite	$10^{-2} - 10^2$
Ilmenite	$10^{-2} - 10^2$
Marcasite	$10^{-2} - 10^2$
Shale	<15
Pyrolusite	$1 - 10$
Coal, Anthracite	$10^{-3} - 5$
Chalcopyrite	$10^{-3} - 10^{-1}$
Sulfides, general	$10^{-3} - 10^{-1}$
Pyrite	$10^{-4} - 10^{-1}$
Magnetite	$10^{-4} - 10^{-2}$
Bornite	$10^{-6} - 10^{-2}$
Galena	$10^{-5} - 10^{-3}$
Pyrrhotite	$10^{-5} - 10^{-4}$
Graphite	$10^{-6} - 10^{-4}$
Metals, Native	$<10^{-6}$

TABLE 3.4a

Densities and Thermal Neutron Capture Cross Sections

Material	Z/A Ratio g/cc	Matrix Density g/cc	Density Range g/cc	Apparent Density g/cc	Cross Section Σ_{ma} 10^{21} b/cc
Lead	0.3953	11.34		8.97	5.61
Uraninite	0.4000	8.25	(6.5–10.8)	6.60	49.69
Cinnabar	0.4143	8.1	(8.0–8.2)	6.71	7981.16
Iron	0.4687	7.87		7.38	214.90
Galena	0.4093	7.5	(7.4–7.6)	6.14	12.47
Wulfenite	0.4187	6.9	(6.7–7.0)	5.78	32.50
Arsenopyrite	0.4605	6.1	(5.9–6.2)	5.62	165.22
Colbaltite	0.4517	6.1	(6.0–6.3)	5.51	936.71

TABLE 3.4a (continued)

Densities and Thermal Neutron Capture Cross Sections

Material	Z/A Ratio g/cc	Matrix Density g/cc	Density Range g/cc	Apparent Density g/cc	Cross Section Σ_{ma} 10^{21} b/cc
Chalcocite	0.4610	5.65	(5.5–5.8)	5.21	173.56
Hematite	0.4787	5.26	(4.9–5.3)	5.04	100.47
Magnetite	0.4774	5.18	(4.97–5.18)	4.95	112.10
Bornite	0.4643	5.15	(4.8–5.4)	4.78	145.63
Pyrite	0.4850	5.06	(4.95–5.17)	4.91	89.06
Illmanite	0.4757	4.75	(4.5–5.0)	4.52	158.23
Zircon	0.4691	4.69	(4.2–4.86)	4.40	5.42
Stibnite	0.4436	4.57	(4.52–4.62)	4.05	17.82
Pyrrhotite	0.4812	4.55	(4.58–4.64)	4.40	90.52
Barite	0.4454	4.45	(4.30–4.60)	3.96	19.40
Chromite	0.4753	4.45	(4.30–4.60)	4.23	102.20
Rutile	0.4756	4.20	(4.15–4.25)	3.80	202.75
Calcopyrite	0.4751	4.2	(4.1–4.3)	3.99	80.95
Corundum	0.4904	4.02	(3.95–4.10)	3.94	11.04
Carnotite	0.4350	4+		3.5	56.21
Rhodocrosite	0.4793	4.0	(3.5–4.0)	3.84	278.93
Sphalerite	0.4720	4.0	(3.9–4.1)	3.78	38.33
Siderite	0.4797	3.88	(3.0–3.88)	3.72	68.81
Limonite	0.4897	3.8	(3.51–4.0)	3.72	74.10
Dunite	0.4978	3.3	(3.24–3.74)	3.29	17.03
Olivine	0.4892	3.3	(3.27–3.37)	3.23	31.74
Magnesite	0.4992	3.1	(3.0–3.2)	3.1	1.48
Norite	0.4970	2.98	(2.72–3.02)	2.97	12.88
Diabase	0.4954	2.98	(2.96–3.05)	2.95	17.21
Gabbro	0.4938	2.98	(2.85–3.12)	2.94	21.47
Anhydrite	0.4995	2.95	(2.89–3.05)	2.95	12.30
Aragonite	0.4995	2.94	(2.85–2.94)	2.94	8.12
Muscovite	0.4966	2.93	(2.76–3.1)	2.91	17.30
Biotite	0.4900	2.90	(2.65–3.1)	2.84	25.20
Dolomite	0.4994	2.85	(2.80–2.99)	2.85	4.78
Illite	0.4954	2.84	(2.60–3.0)	2.81	39.90
Diorite	0.4964	2.84	(2.72–2.96)	2.82	14.33
Langbeinite	0.4961	2.83		2.61	78.87
Polyhalite	0.5013	2.78		2.79	21.00
Synite	0.4971	2.75	(2.630–2.90)	2.74	16.43
Granodiorite	0.4963	2.71	(2.67–2.78)	2.696	11.33
Chlorite	0.5056	2.71	(2.60–3.22)	2.74	17.56
Calcite	0.4996	2.71	(2.71–2.72)	2.71	7.48
Aluminum	0.4818	2.70		2.60	13.99
Plagioclase Feldspar	0.4925	2.69	(2.62–2.76)	2.65	6.99
Limestone	0.5000	2.69	(2.66–2.74)	2.69	8.72
Granite	0.4969	2.67	(2.52–2.81)	2.65	11.62
Quartz	0.4993	2.65	(2.65–2.66)	2.65	4.36
Sandstone	0.4990	2.65	(2.59–2.84)	2.655	8.66

Room temperature; one atmosphere; averages of multiple samples.

TABLE 3.4b

Densities and Neutron Capture Cross Sections

Material	Z/A Ratio	Matrix Density g/cc	Density Range g/cc	Apparent Density g/cc	Σ_{ma} c.u. $10^{21}b/cm^3$
Kaolinite	0.5103	2.63	(2.40–2.68)	2.68	13.06
Albite	0.4885	2.62	(2.61–2.65)	2.56	6.77
Orthoclase Feldspar	0.4958	2.57	(2.55–2.63)	2.55	16.00
Kieserite	0.4724	2.57		2.43	12.77
Concrete		2.35	(1.98–2.35)		
Montmorillinite	0.5009	2.35	(2.00–3.00)	2.35	8.10
Gypsum	0.5111	2.32	(2.30–2.35)	2.37	19.40
Glauconite	0.4998	2.30	(2.20–2.80)	2.30	16.80
Graphite	0.4995	2.22	(2.09–2.23)	2.22	0.38
Serpentine	0.5062	2.20		2.23	8.80
Halite	0.4799	2.16	(2.135–2.165)	2.07	752.36
Nahcolite	0.4905	2.20		2.16	
Kainite	0.5140	2.13	(2.1–2.13)	2.19	196.13
Trona	0.5043	2.12	(2.11–2.15)	2.14	16.21
Sulfur, Orthorhombic	0.4990	2.07	(2.05–2.09)	2.07	19.06
Potash	0.5049	2.04		2.06	39.70
Sylvite	0.4829	1.99	(1.97–1.99)	1.92	570.68
Cement		1.99			~13
Sulfur, Monoclinic	0.4990	1.96		1.96	18.05
Kernite	0.5049	1.91		1.92	12793.69
Carnalite	0.4829	1.61	(1.60–1.61)	1.64	370.92
Coal, Anthracite	0.5134	1.60	(1.32–1.80)	1.64	1.08
Coal, Bituminous	0.5201	1.35	(1.15–1.7)	1.40	1.54
Coal, Lignite		1.10	(0.5–1.5)	1.16	
Water (3×10^5)	0.5325	1.219		1.298	146.22
(2.5×10^5)	0.5363	1.1825		1.268	122.55
(2×10^5)	0.5401	1.146		1.238	100.08
(1.5×10^5)	0.5438	1.109		1.206	78.75
(10^5)	0.5476	1.073		1.175	58.69
(5×10^4)	0.5513	1.0365		1.143	39.02
(3×10^4)	0.5528	1.022		1.130	32.56
(0ppm)	0.5551	1.00		1.11	22.08
Petroleum (10°API)	0.5703	1.00		1.14	28.02
(30°API)	0.88	1.00			25.89
(40°API)	0.85	0.97			24.22
(50°API)	0.78	0.85			22.23
(70°API)	0.5778	0.70		0.81	22.12
N-Pentane	0.5823	0.626		0.733	20.80
N-Hexane	0.5803	0.659		0.765	21.38
N-Heptane	0.5778	0.684		0.790	21.80
N-Octane	0.5778	0.703		0.812	22.12
N-Nonane	0.5768	0.718		0.828	22.37
N-Decane	0.5763	0.730		0.841	22.55
N-Undecane	0.5759	0.740		0.852	22.71
Methane	0.5703	0.000677		0.00076	0.028
Ethane	0.5986	0.001269		0.00015	0.051
Propane	0.5896	0.00186		0.0022	0.067
N-Butane	0.5850	0.00246		0.0029	0.085

TABLE 3.4b (continued)

Densities and Neutron Capture Cross Sections

Material	Z/A Ratio	Matrix Density g/cc	Density Range g/cc	Apparent Density g/cc	Σ_{ma} c.u. 10^{21}b/cm^3
Helium	0.4997	0.00017		0.00017	0.0000
Carbon dioxide	0.4998	0.001858		0.001857	0.0001

Water values in ppm NaCl; all samples at room temperature, one atmosphere, averages of multiple samples

TABLE 3.4c

Element Densities and Neutron Capture Cross Sections

Material	Z/A Ratio g/cc	Matrix Density g/cc	Σ_{ma} 10^{21} b/cc
Nitrogen	0.4998	0.001182	0.004
Oxygen	0.5000	0.001350	0.00001
Hydrogen sulfide	0.5281	0.001438	0.029
Air (dry)	0.4997	0.001224	
Argon	0.4859	0.001688	0.017
Natural gas, aver.	0.5735	0.0007726	
200°F, 7000psi	0.5735	0.252	
Hydrogen	0.9921	0.00009	
Carbon	0.4995	3.52	
Calcium	0.4990	1.5	
Silicon	0.4985	2.4	
Magnesium	0.4975	1.75	
Potassium	0.4859	0.86	
Phosphorous	0.4845	1.83	
Sodium	0.4804	0.97	
Chlorine	0.4795	0.0032	
Nickel	0.4769	8.90	
Iron	0.4687	7.86	
Boron	0.4625	2.45	
Chromium	0.4614	7.1	
Titanium	0.4593	4.5	
Zinc	0.4589	7.14	
Manganese	0.4568	7.4	
Copper	0.4564	8.92	
Vanadium	0.4514	5.96	
Cobalt	0.4432	8.9	
Arsenic	0.4405	5.7	
Zirconium	0.4385	6.4	
Bromine	0.4380	3.12	
Strontium	0.4336	2.6	
Tin	0.4212	7.2	
Molybdenum	0.4159	10.2	
Barium	0.4077	3.5	
Mercury	0.3988	13.56	
Uranium	0.3865	18.7	

Room temperature; one atmosphere; averages of multiple samples.

TABLE 3.5

Potassium Content of Formation Materials

Material	Potassium Content by Weight %	Range
Sylvite	54	
Potash	44.9	
Lanbeinite	20	
Microcline	16	
Kainite	15.1	
Carnalite	14.1	
Orthoclase	14	
Polyhalite	12.9	
Muscovite	9.8	
Biotite	8.7	
Illite	5.2	3.51–8.31
Arkose (Sandstone)	4.6	4.4–5.1
Syenite	4.53	
Glauconite	4.5	3.2–5.8
Granite	4.0	2.0–6.0
Norite	3.3	
Granodiorite	2.90	
Shale	2.7	1.6–9.0
Igneous Rock	2.6	
Graywacke (Sandstone)	1.8	1.2–2.1
Diorite	1.66	
Basalt	1.3	
Sandstone (Not clean)	1.1	0–5.1
Gabbro	0.87	
Diabase	0.75	
Kaolinite	0.63	0–1.49
Limestone	0.27	0–0.71
Montmorillinite	0.22	0–0.60
Orthoquartzite (Sandstone)	0.08	0–0.12
Dolomite	0.07	0.03–0.1
Dunite	0.04	
Sea Water	0.035	

TABLE 3.6

Approximate Acoustic Travel Times and Velocities

Material	Matrix Travel Time µs/ft		Matrix Velocity ft/s	
	Average	Range	Average	Range
Dunite	38.2	34.7–41.1	26,174	24,305–28,807
Gabbro	42.4	42.2–47.6	23,586	20,998–23,586
Hematite		42.9		23,295
Dolomite	44.0	40.0–45.0	22,727	22,222–25,000
Norite	44.1	43.5–49.0	21,683	20,400–22,967

TABLE 3.6 (continued)

Approximate Acoustic Travel Times and Velocities

Material	Matrix Travel Time µs/ft		Matrix Velocity ft/s	
	Average	Range	Average	Range
Diabase	44.6	44.0–46.0	22,435	21,746–22,730
Anorthosite		45.4		22,016
Calcite	46.5	45.5–47.5	21,505	21,053–22,000
Aluminum		48.7		20,539
Anhydrite		50.0		20,000
Albitite	50.2	49.5–50.6	19,916	19,752–20,212
Granite	50.8	46.8–53.5	19,685	18,691–21,367
Granite		50.8		19,700
Steel		50.8		19,686
Limestone	52.0	47.7–53.0	19,231	18,750–21,000
Langbeinite		52.0		19,231
Iron		52.1		19,199
Gypsum	53.0	52.5–53.0	19,047	18,868–19,047
Serpentine		53.9		18,702
Quartzite	55.0	52.5–57.5	18,182	17,390–19,030
Quartz	55.1	54.7–55.5	18,149	18,000–18,275
Sandstone	57.0	53.8–100	17,544	10,000–19,500
Casing, steel		57.1		17,500
Basalt		57.5		17,391
Polyhalite		57.5		17,391
Shale		60.0		5,882–16,667
Aluminum tube		60.9		16,400
Trona		65.0		15,400
Halite		66.7		15,000
Sylvite		74.0		13,500
Copper		78.7		12,700
Carnalite		83.3		12,000
Cement,	95.0	83.3–95.1	10,526	10,526–12,000
Coal, Anthracite	105.0	90.0–120.0	9,524	8,333–11,111
Concrete	95.2	83.3–125.0	10,500	8,000–12,000
Coal, Bituminous	120.0	100.0–140.0	8,333	5,906–10,000
Sulfur		122.0		8,200
Coal, Lignite	160.0	140.0–180.0	6,250	3,281–7,143
Lead		141.1		7,087
Water, 2 × 10⁵	180.5		5,540	
1.5 × 10⁵	186.0		5,537	
10⁵		192.3		5,200
Pure		207.0		4,830
Glacial Ice		87.1		11,480
Rubber		190.5		5,248
Kerosene		214.5		4,659
Oil		238.0		4,200
Methane		626.0		1,600
Air		910.0		1,100

Water values are in ppm of NaCl.
Fluids were measured at 15 psia.

TABLE 3.7

Thermal Conductivities of Various Materials

Material	Thermal Conductivity	Temp (°F)	Material	Thermal Conductivity	Temp (°F)
Carbon Dioxide	0.040	80	Shale	2–40	
	.044	120	Sandstone	3–12.2	
Ethane	0.043	32	Fused Silica	3.2	
	0.051	80	Limestone	2.4–8	
	0.074	200	Polyhalite	3.7	
Air, dry	0.057	32	Serpentine	4.3–5.9	
	0.061	68	Basalt	4–7	
	0.074	212	Granite	5–8.4	
	0.088	392	Calcite	5–13	
Nitrogen	0.062	80	Feldspar	5.8	
	0.065	120	Slate	6	
Methane	0.073	32	Norite	6.42	
	0.081	80	Granodiorite	6.64	
	0.106	200	Quartz	6–30	
Crude Oil	0.3	68	C-axis	26	100
Helium	0.332	68		22	200
Kerosene	0.357	86	A-axis	14	100
Sulfur, momoclinic	0.38	212		12	200
Sulfur, rhombic	0.56	176	Syenite	7.66	
	0.65	68	Salt	14.3 (8–15)	
Coal, Lignite	0.33–1	68	Garnet	8.5	
Cement, Portland	0.71	68	Dolomite	9.3–11.9	
Water	1.39	32	Dunite	10	
	1.43	68	Chlorite	12.5	
	1.60	167	Anhydrite	7–13.4	
Clay	2–3		Quartzite	16.05	
Chalk	2–3		Pyrite	25–40	
Gypsum	3.6 (2–4)		Hematite	25	
Lead	83		Magnetite	30	
Iron	180	68	Sphalerite	63.6	
Magnesium	380	68			
Aluminum	530	68			
Copper	940	68			

Thermal Conductivities units are: $10^{-3} \dfrac{cal/cm^2\ sec}{°C/cm}$

TABLE 3.8

Magnetic Susceptibilities

Material	K 10^4 C.G.S.	H Oersteds
Magnetite	$3 \times 10^5 – 8 \times 10^5$	0.6
Ilmenite	1.35×10^5	1
Pyrrhotite	1.25×10^5	0.5
Franklinite	3.6×10^4	—
Serpentine	1.4×10^4	30.5
Peridotite	1.25×10^4	0.5–1.0
Olivine, Diabase	2×10^3	0.5
Granite	$27–2.7 \times 10^3$	1
Gabbro	$67–2.37 \times 10^3$	1
Diabase	$77–1.05 \times 10^3$	1
Basalt	680	1
Diorite	47	1
Porphyry	47	1
Sandstone	17	1
Dolomite	14	0.5
Sulfur, Rhombic	4.7×10^{-1}	
Sulfur, Monoclinic	4.7×10^{-1}	

Sulfur measured at 18°C

TABLE 3.9

Solubilities of Minerals in Water

Material	°Celcius	Solubility Cold Water g/100cc	°Celcius	Solubility Hot Water g/100cc
Kieserite		n/a		68.4 [7]
Aluminum		i		i
Chromite		i		i
Graphite		i		i
Hematite		i		i
Iron		i		i
Lead		i		i
Limonite		i		i
Magnetite		i		i
Quartz		i		i
Rutile		i		i
Sulfur, Orthorhombic		i		i
Sulfur, <95.4°C		i		i
Uraninite		i		i
Zircon		i		i
Chalcocite	18	10^{-14} [1]		n/a
Cinnabar	18	10^{-6} [1]		n/a
Galena	20	8.6×10^{-5} [3]		n/a
Corundum	18	9.8×10^{-5} [1]		n/a
Stibnite	25	1.75×10^{-4} [1]		n/a
Barite	25	2.22×10^{-4}	75	3.36×10^{-4} [8]
Pyrite	25	4.9×10^{-4}		n/a
Sphalerite	18	7×10^{-4} [1]		n/a
Calcite	25	1.4×10^{-3} [4]	75	1.8×10^{-3} [9]
Rhodochrosite	20	2.65×10^{-3} [3]		n/a
Siderite	25	6.7×10^{-3} [4]		n/a
Magnesite	19	1.06×10^{-2}		n/a
Dolomite	18	3.2×10^{-2} [1]		n/a
Aragonite	25	2.00×10^{-1} [4]	75	1.90×10^{-3} [9]
Anhydrite	30	2.09×10^{-1} [5]	100	1.619×10^{-1} [7]
Gypsum		2.41×10^{-1}	100	2.22×10^{-1} [7]
Nahcolite	0	6.9 [6]	60	16.4 [10]
Trona	0	13.0 [6]	100	42.0 [7]
Sylvite	0	28.0 [6]	100	56.0 [7]
Halite	0	35.6 [6]	100	39.8 [7]
Carnalite	19	64.5 [2]		n/a
Potash	20	110.5 [3]	100	155.7 [7]

[1] 18°C   [6] 0°C
[2] 19°C   [7] 100°C
[3] 20°C   [8] 50°C
[4] 25°C   [9] 75°C
[5] 30°C   [10] 60°C

TABLE 3.10

The Hydrogen Content of Various Formation Materials

Material	Temperature °F	Pressure psi	Hydrogen Atoms ×10²³ per cc	Index
Water, pure	60	14.7	0.669	1.0
	200	7k	0.667	1.0
Water, salt,	60	14.7	0.614	0.92
	200	7k	0.602	0.90
Methane,	60	14.7	0.0010	0.0015
	141	4k	0.275	0.41
	200	7k	0.329	0.49
Ethane,	60	14.7	0.0015	0.0023
	200	7k	0.493	0.74
Natural gas	60	14.7	0.0011	0.0017
	200	7k	0.363	0.54
N-pentane	68	14.7	0.627	0.94
	200	7k	0.604	0.90
N-hexane	68	14.7	0.645	0.96
	200	7k	0.615	0.92
N-heptane	68	14.7	0.658	0.99
	200	7k	0.632	0.95
N-octane	68	14.7	0.667	1.00
	200	7k	0.639	0.96
N-nonane	68	14.7	0.675	1.01
	200	7k	0.645	0.97
N-decane	68	14.7	0.680	1.02
	200	7k	0.653	0.98
N-undecane	68	14.7	0.684	1.02
	200	7k	0.662	0.99
Bituminous coal			0.442	0.66
Carnalite			0.419	0.63
Limonite			0.369	0.55
Cement			0.334	0.50
Kernite			0.337	0.50
Gypsum			0.325	0.49
Kainite			0.309	0.46
Trona			0.284	0.42
Potash			0.282	0.42
Anthracite coal			0.268	0.40
Kaolinite			0.250	0.37
Chlorite			0.213	0.32
Kieserite			0.210	0.31
Serpentine			0.192	0.29
Nahcolite			0.158	0.24
Glauconite			0.127	0.19
Montmorillinite			0.115	0.17
Polyhalite			0.111	0.17
Muscovite			0.089	0.13
Illite			0.059	0.09
Biotite			0.041	0.06

The "Hydrogen Index" is the equivalent neutron porosity.

4

Porosity, Permeability, Tortuosity, and Saturation

4.1 General

Porosity is important in petroleum work because it is the space in reservoirs in which the water, gas, and hydrocarbons are found. In sedimentary mineral deposits, the porosity is the space through which the mineralizing solutions move. The amount of water in the pore space partially determines the quality of a hydrocarbon deposit and also must be accounted for before a mine can be built. Porosity (in rock) is a relative measure of the non-rock space within a formation material. It is a fraction of the bulk volume of the zone.

Tortuosity describes the shape of the interconnected pore space. It is important in the measurement of resistivity and affects the permeability. Tortuosity implies porosity.

Permeability is the ability of fluids to move through the rock (or other) material. Permeability implies porosity and bears a relationship to porosity. On the other hand, porosity can exist in substantial amounts without permeability.

Saturation is the relative amount of fluid within the interconnected pore space. It is a measure of the relative amount of water, hydrocarbon, and other fluids in the pore space. Saturation implies porosity. Note that saturation is a fraction of the pore space.

4.2 Porosity

Porosity is the fraction of the volume of the void space, or the interstices within the rock body. It is the complement of the rock matrix. Porosity value is always given as a unitless fraction or percent of the total rock volume. The pore space will always be filled with fluid and/or frequently with a shale. Pore spaces usually represent a substantial contrast in physical properties with respect to the rock material. We exploit this in geophysical logging.

We are particularly interested in the porosity in petroleum work because it is where the hydrocarbon will be found in producible quantities. It is where the water is found for hydrological work . Permeability is also needed. In mineral work, the interest lies in the rock portion. The interest may lie in either material or both in engineering and scientific work. Porosity and permeability are of particular interest in engineering work. The amounts and salinities of ground waters are interesting in environmental concerns.

In the late 19th century, it was believed that the overburden pressure collapsed the pore space below about 1000 meters depth and that porosity did not exist below that depth. We know now, of course, that this is not true. Evidence of porosity below 35,000 feet (10,000 meters) is common. We do have evidence, however, of decreasing porosity with depth (Figure 4.1).

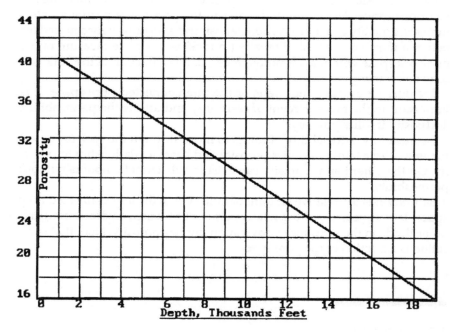

FIGURE 4.1
Average Texas Gulf Coast clean sand apparent porosity vs. formation depth.

Core porosities, measured in the laboratory, are usually accepted as the standard, against which other methods are compared. This does not mean that core determinations are infallible. It means merely, that the core laboratory methods are more easily controlled. Also, historically, core methods predate wireline methods.

There are several kinds of porosity which we will consider. Total porosity is all of the nonrock fraction of the total or bulk volume of the rock sample. It includes the space occupied by clay or shale, liquid, and/or gas. It also includes the nonpermeable or isolated pore space. Effective porosity is that fraction of the total rock volume available to the free flow of fluid. This is the interconnected porosity. It is typical in a clean sand.

Isolated porosity is non-connected void space. It is part of the total porosity. It may be found in carbonates, highly cemented zones, and in igneous rocks. It is generally nonproductive for petroleum purposes. It becomes very important in engineering work. Some of the geophysical methods (i.e., electrical resistivity) will not respond to isolated porosity in most circumstances. Some projects have successfully tried producing the isolated porosity for gas. The nuclear stimulation test at the Green River location in Wyoming and Colorado is an example.

The difference between total porosity and effective porosity is that volume fraction occupied by bound water, shale, and isolated porosity. (Shale volume, V_{sh}, is the fraction of the bulk volume occupied by clay or shale.)

Primary porosity is developed and modified during deposition, compaction and cementing. It depends upon grain shape and arrangement, angularity, and grain size distribution. It is typical in a clean, unconsolidated sand. This is frequently intergranular porosity. Secondary porosity is developed after deposition and compaction. It is due to leaching, fracturing, shrinkage. Its forms are vugular, solution cavities (big vugs), and fractures. It is typical in many carbonates. It is found in fractured, altered, and faulted zones.

Porosity increases for greater angularity and lower distribution of grain size (better sorting). The range is usually from 10% to 35% of the bulk volume, but may be lower. The lower porosity ranges are found in older rocks and/or deeper rocks. This is due to weathering, compaction, and cementing. This is the probable basis for a discovery by George Keller of the Colorado School of Mines, that the average resistivity, after temperature corrections, increases with the age of the formation.

The fluid in the pore spaces usually is water/salt solution: the native water (formation water, connate water), mud filtrate or other drilling fluid. It may also be hydrocarbon liquid, hydrocarbon gas, non-hydrocarbon gas, or a mixture of any or all of these.

4.3 Tortuosity

Within the rock, the rock material (i.e., quartz) has extremely high electrical resistivity. The amount of electrical current is inversely proportional to the resistivity, for a given voltage difference. In rock materials, electrical resistivity will usually be on the order of 10^{10} to 10^{17} times that of the conducting materials. In contrast, water has a resistivity range from 2×10^{-2} to 50 ohmmeters. Metallic conductors have resistivities less than 10^{-6} ohmmeters.

Thus, since the rock has such a high resistivity compared to the water filling the pore spaces, virtually all of the current will flow in the water. As far
as our logging tools are concerned, the rock material is essentially noncon-
ductive. For our purposes, we will consider that rock materials are perfect
insulators. Thus, in any rock, and especially in sediments, virtually all
electrical currents are carried by the ions in the water; in the voids, the
pores, fractures, vugs, etc. These relationships are of major importance in
petroleum work, because the presence of hydrocarbon or gas (10^9 to 10^{14}
ohmmeters) in the pore spaces will reduce the apparent volume of the pore
space, as seen by the resistivity device, by an amount equal to $(1 - S_w)$.

Until recently, no other measurement family had the sensitivity to the
presence of hydrocarbon that the resistivity systems have. The pulsed neutron systems and the electromagnetic propagation time (EPT) devices now
approach the resistivity systems in this respect.

The character of the formation fluid (the salinity, temperature, type
fluid, etc.) affects the log response. The lithology or rock type will affect
many logs to one degree or another, but not the resistivity and conductivity logs. We have seen that the amount of porosity will affect most logs.

For a particular rock sample which contains interconnected pore spaces
(i.e., sandstones) and which is 100% saturated with electrolyte (water), picture a block of material composed of rods the length of the block
(Figure 4.2a and b). Each rod has a unit cross sectional area, a, and a length,
L. The cross section of the block, $A_{total} = \Sigma a$. Since the resistance of any
material is shown by:

$$r = R\frac{L}{A} \tag{4.1}$$

The cross-sectional area and length of an single electrolyte path, which can
represent the sum of all of the paths in that sample, can be described as:

$$\frac{L}{A} = \frac{r}{R_w} \tag{4.2}$$

where

    L = the path length
    A = the path cross sectional area
    r = the resistance, in ohms, of the path
    R_w = the resistivity, in ohmmeters, of the electrolyte filling the path

The resistance, r, is determined solely by the geometry of the pore spaces
and the resistivity of the fluid filling them. Thus, the "geometric ratio" is

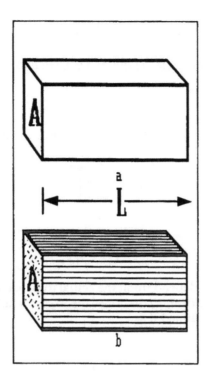

FIGURE 4.2
A hypothetical box for resistivity, porosity, tortuosity demonstration.

$$\frac{L_e}{A_e} = \frac{r}{R_w} \tag{4.3}$$

where R_w is the resistivity of the saturating fluid and the subscript "e" denotes that these values are equivalents representing the electrolytic path. Thus, L_e and A_e are the effective length and area of the electrolyte. The value of the geometrical ratio, L_e/A_e, can be determined because the values of r and R_w can be measured readily. The values of L_e and A_e can also be determined.

The porosity, ϕ, of the sample is the ratio of the total absolute volume of the pore space, V_p, to the bulk volume of the sample, V_b:

$$\phi = \frac{V_p}{V_b} \tag{4.4}$$

If the sample contains no isolated porosity and only interconnected pore spaces (i.e., sandstone) then:

$$V_p = L_e A_e \tag{4.5}$$

Substituting Equation 4.4 into Equation 4.3, we find:

$$L_e A_e = \phi V_b \tag{4.6}$$

Combining with Equation 4.3 and solving for L_e and A_e, we find:

$$L_e = \left[\frac{\phi r V_b}{R_w}\right]^{1/2} \tag{4.7}$$

$$A_e = \left[\frac{\phi V_b R_w}{r}\right]^{1/2} \tag{4.8}$$

L_e and A_e, of course, depend upon the actual dimensions of the sample.

The *tortuosity coefficient*, t_p, describes the excess length of L_e over that of the sample. Pirson defined it as:

$$t_p = \frac{L_e}{L_b} \tag{4.9}$$

where L_b is the bulk length of the sample.

The *diminution coefficient*, *d*, describes the constriction of the cross-sectional area, A_e of the equivalent pore. This was defined by Frasier and Ward as:

$$d = \frac{A_e}{A_b} \tag{4.10}$$

Archie proposed a "formation resistivity factor", F_r, to be:

$$F_r = \frac{R_o}{R_w} \tag{4.11}$$

Frasier and Ward showed the F_r is related to the tortuosity:

$$F_r = \frac{t}{d} \tag{4.12}$$

and further, that:

$$t = \left(F_r \phi\right)^{1/2} \tag{4.13}$$

$$d = \frac{\phi^{-1/2}}{F_r} \tag{4.14}$$

For a further discussion of porosity and tortuosity, especially as they affect the resistivity measurement, refer to Chapter 7. Also refer to Doveton, 1986.

4.4 Permeability

Permeability is the measure of the ease with which fluids can pass through the interconnected pore spaces of the formation material without altering the rock matrix. It is a measure of the fluid conductivity of the rock. Its unit is the "darcy", named for Henry Darcy. It is defined as the capacity of a material to transmit or conduct fluids. The permeability of one darcy will allow the flow of one milliliter per second of a single phase fluid of one centipoise viscosity which completely fills the voids, through one square centimeter of a medium, under a pressure gradient of one atmosphere per centimeter. Permeability, k, must exist for the fluid to be producible. The unit of measure of permeability, is the darcy; but the millidarcy is commonly used. Darcy's Law states:

$$q = \frac{kA}{\mu}\frac{dp}{dx} \tag{4.15}$$

where

q = the volume per unit time (cm^3/sec)

k = the permeability constant in darcys

A = the cross sectional area of flow (cm^2)

μ = the fluid viscosity (centipoise)

dp/dx = the hydraulic gradient (atmospheres per centimeter)

Equation 4.15 assumes that the flow of fluid is laminar or viscous and is homogeneous. It further assumes that the transmitting medium has a uniform packing and cross section. These assumptions, of course, are quite simplified. Natural fluids are seldom single phase nor uniform nor are pore systems uniform. Most of the fluids we will work with may be single, two, or three phase. That is, they may be composed of any combination of liquids which are immiscible or miscible and/or gases. Thus, the effective

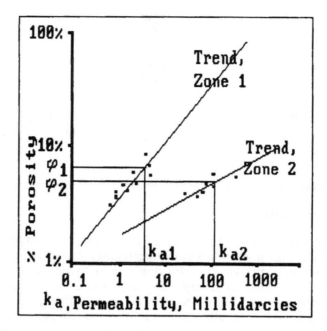

FIGURE 4.3
Plotting of permeability vs. porosity can show trends of rock characteristics.

permeability, the ability of a rock to conduct a fluid in the presence of other fluids, is important. Relative permeability is the ratio of the ability to conduct an actual fluid to its ability to conduct 100% water. The ability to conduct a single phase fluid, such as water or a gas, is the absolute permeability.

Permeability can be measured on core samples in the laboratory, or, it can be estimated or measured with wire-line and formation-testing methods. Permeabilities measured from the core samples are much more reliable, generally, than estimates or calculations from logged data.

Porosity and permeability are related. Permeability is a function of the porosity and of the pore shape (the tortuosity). See Figure 4.3. The permeabilities of sands may be plotted against their porosity, if the tortuosity is constant. Very frequently we can find an approximate relationship between the porosity and the permeability of a zone. Some of the common ones are:

$$k_1 = A_1 \frac{\phi^3}{f(1-\phi)^2 S^2} \tag{4.16}$$

$$k_2 = A_2 \frac{\phi}{(1-\phi)^2 S_o^2} \tag{4.17}$$

$$k_3 = A_3 \frac{\phi}{S_p^2} \qquad (4.18)$$

$$k_4 = B_4^{(-B\phi)} \qquad (4.19)$$

where

S = the inner surface of the rock per unit bulk volume, in cm²/cm³

A_1 = an empirical (Kozeny's) constant

A_2 = an empirical constant

S_o = the surface area per unit volume of solid material (the rock matrix)

S_p = the pore surface area per unit porosity

A_3 = an empirical constant

S_p = the surface area per unit volume of pore space

B and B_4 = empirical constants

As you may gather from this, the estimation of permeability from only the pore space value, is rather a matter of experience. The inclusion of the irreducible oil saturation value, S_{ir}, in the permeability estimate tends to increase the reliability of the estimated value. This is because the permeability and the irreducible oil saturation are both related to the capillary pressure in the porous medium. The expression suggested by Tixier, Timur, Morris, Biggs, Coates, Dumanoir among others was:

$$k_5 = A_5 \frac{\phi^B}{(S_{wi})^{C_5}} \qquad (4.20)$$

where S_{wi} is the irreducible water saturation and A_5, B_5, and C_5 are empirical constants. Timur suggested $A_5 = 0.136$, $B_5 = 4.4$, and $C_5 = 2$. Morris and Biggs suggested $A_5 = (250)^2$ for medium gravity oil and $(79)^2$ for dry gas, while $B_5 = 6$, and $C_5 = 2$. Coates and Dumanoir suggested $A_5 = 10^4$, $B_5 = 4.5$, and $C_5 = 2$. Again, these relationships were mostly empirical. They have, however, served (albeit poorly) for many years. Note that the values of the constants vary widely; these are relationships each of which was adopted to suit the conditions of a specific area. A given relationship may serve poorly in other places. This, however, is typical of many of the empirical approaches used in the oilfield.

Further refinements of this approach were suggested by Schlumberger:

$$k_6 = C_6 a \left[\frac{2.3}{\rho_w - \rho_h} \right]^2 \qquad (4.21)$$

$$a = \frac{\Delta R}{\Delta D} \frac{1}{R_o} \qquad (4.22)$$

where

C_6 = a constant, which is normally about 20
ΔR = the change in resistivity
ΔD = the change in depth corresponding to
R = the resistivity of the formation material
R_o = the 100% water saturated formation resistivity
ρ_w = the formation water density
ρ_h = the density of the hydrocarbon

The assumption is that a granular structure and an irreducible saturation are present. The obvious problem with this method is that the zone is at irreducible saturation. This method, however, was based mostly upon laboratory experiments and not upon log readings.

Further suggestions, by Jones, were:

$$k_{rw} = \left[\frac{S_w - (S_{wi})^3}{1 - S_{wi}} \right]^3 \qquad (4.23)$$

and:

$$k_7 = \frac{A^E}{(S_w)^E (P_F)^F} \qquad (4.24)$$

By Brown and Hussein, where:

Sw = the water saturation
k_{rw} = the relative permeability to water
P_G = the capillary pressure
A_6 = 57
E = 0.86
G = 1.26
F = 0.89

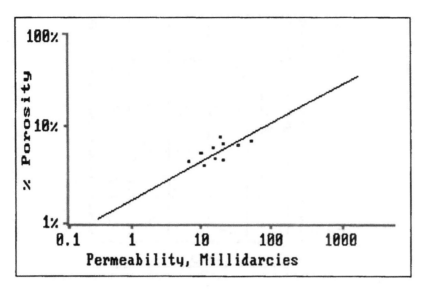

FIGURE 4.4
A crossplot for predicting probable porosities and/or permeabilities in a single horizon.

Brown and Hussein also suggested that A_6 could equal 22.8, $E = 1.38$, and $G = 1.05$. This was found to produce a value for permeability about equal to the previous equation and could be used in zones with intermixed lithologies. Equations 4.23 and 4.24 also were developed empirically from logged data.

Success has been achieved in measuring the permeability *in situ* with acoustic techniques. Refer to a text on acoustic logging for a more detailed discussion of these techniques. Simply plotting core permeabilities against the porosity is a common technique called cross plotting. We will use it often. See Figure 4.4. Very few of these diagrams consider tortuosity, therefore, they are sometimes difficult to apply. The problem is that the tortuosity can change as the porosity changes, due to cementation and fracturing.

Sedimentary formation material permeabilities normally range from 0.01 millidarcy to about 20,000 md. The accuracy of laboratory measurement of this declines at the upper and lower ends, but generally is about ±5%. Permeability is constant when,

1. No reaction occurs between the rock and the flowing fluid
2. Laminar flow exists
3. A single fluid completely saturates the core

Laboratory measurement of permeability generally uses a one phase fluid and is the absolute permeability, k_a. It is measured by flowing a gas (often helium) through a clean, dry plug from a core.

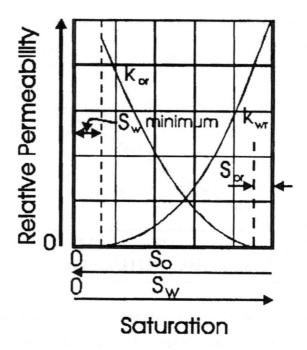

FIGURE 4.5

The approximate relationship between permeability and saturation. (From Amyx, J. et al., *Petroleum Reservoir Engineering: Physical Properties*, McGraw-Hill, 1960. With permission.)

When we have a two phase flow (a mixture of 2 fluids) such as oil and water, we find that there is a permeability, k_o, for the water and one, k_o, for the oil and that they are not the same. The relative permeability of the oil, K_{or},

$$k_{or} = \frac{k_o}{k_a} \qquad (4.25)$$

Similarly, the relative permeability of the water, k_{wr}, is:

$$k_{wr} = \frac{k_w}{k_a} \qquad (4.26)$$

The relationship between them is shown in Figure 4.5.

When the permeable medium has two or more different types of fluids, then the permeability becomes a function of the saturations, as well as of the porosity and tortuosity.

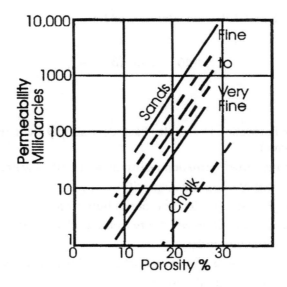

FIGURE 4.6
A plot of porosity vs. permeability, as a function of grain size. (From Master's thesis Courtesy of H. Vasquez, 1961.)

4.5 Saturation

About 1940, Gerald Archie of the Shell Oil Company determined empirically:

$$F_r = \frac{R_o}{R_w} = \frac{R_{xo}}{R_{mf}} \qquad (4.27)$$

when water saturation is 100%,

$$F_r = \phi^{-2} \qquad (4.28)$$

$$I_r = \frac{R_t}{R_o} \qquad (4.29)$$

$$\left(S_w\right)^{-2} = I_r \qquad (4.30)$$

Equations 4.28 and 4.30 were later revised to:

$$F_r = a\phi^{-m} \quad \text{and} \quad \left(S_w\right)^{-n} = I_r \qquad \text{(4.31a, b)}$$

where:

F_r = the Formation Resistivity Factor (or simply the Formation Factor)

R_o = the measured formation resistivity when 100% water saturated

R_w = the formation water resistivity

ϕ = the porosity as a fraction of the bulk formation volume

R_t = the real formation resistivity in the undisturbed formation

I_r = the resistivity index

S_w = the water saturation as a fraction of the pore space

m = the "cementation" exponent

n = the saturation exponent

The factor, a, is an empirical constant which is usually assigned a value of 1.0. It was added to force the Archie equation to conform to the effects of tortuosity. A notable example of this, latter, is the "Humble" relationship, which can be used for soft and unconsolidated sands:

$$F_{r,H} = 0.62\,\phi^{-2.15} \qquad \text{(4.32)}$$

The Humble relationship is *frequently* and *incorrectly* used as a "universal" relationship. It is much safer and more nearly correct to use one of the newer relationships, even in unconsolidated sands. The value of "a" will be assumed to be 1.0 throughout this text, unless stated otherwise.

The water saturation, S_w, is the fraction of the pore space filled with water. Oil saturation, S_o, if no gas is present, is

$$S_o = 1 - S_w \qquad \text{(4.33)}$$

Similarly for the gas saturation, S_g:

$$S_g = 1 - S_w \qquad \text{(4.34)}$$

Obviously, the general relationship for hydrocarbon saturation, S_h, which does not distinguish between the types of hydrocarbon, is better:

$$S_h = 1 - S_w \tag{4.35}$$

The irreducible water saturation is S_{wirr}, or simply S_{wi}. We will not be able to remove water below that value, except by heating and drying the material. The water saturation in the flushed zone where invasion has taken place is S_{xo}. The residual oil saturation is S_{or} (or S_{gr}). It indicates the amount of oil (or gas) left in the material by normal production methods. It is equal to or very nearly equal to:

$$S_{or} = 1 - S_{xo} \tag{4.36}$$

S_c is the cutoff value of the water saturation. This is the highest value of water saturation which can be economically tolerated. And by implication, $(1 - S_c)$ is that of the hydrocarbon, S_h. These are *economic values*. Their values will depend upon how much it costs to produce the zone and transport the hydrocarbon and its selling price.

A plot of permeability vs. porosity enables us to predict the probable permeability, from a log value where a core value does not exist. Since there is a separate trend for each type of rock, this particular plot would enable us to evaluate the rock character (i.e., grain size sorting) in the zones where they are not known. A version of this type of plot is shown in Figure 4.4. When this type of plot is used, one must have some information about the formation and a means (core analysis) to check some of the values. Physically, saturation may take several forms. These are illustrated in Figure 4.8. When a zone has hydrocarbon at the top and there is evidence that it has migrated upward from lower in the zone, leaving a lower portion 100% water saturated, it is possible to estimate several things. This will be covered in more detail in Chapter 7. One thing that can be determined is the approximate water saturation, from the rate of increase of the formation density and the resistivity, as one moves from the R_o (100% water saturated) zone upward into the minimum water saturated zone. This is illustrated in a log in Figure 4.9 and will be discussed further in Chapter 7. A chart illustrating this is shown as Figure 4.10.

4.6 Overpressure

As a zone, in time, becomes buried deeper and deeper, the solid-rock matrix collapses slightly. The liquid portion is squeezed out into the adjacent, more permeable zones. The overburden continues to be supported by the rock matrix. The overburden pressure will depend upon the density of the overburden rock. This will be approximately 2.5 to 2.9 g/cc. The

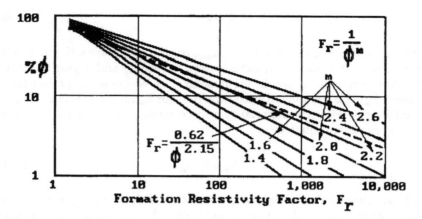

FIGURE 4.7
A plot of the porosity, ϕ, vs. the formation resistivity factor, F_R, as a function of the cementation exponent, m.

pressure of the liquid within the zone will be that of the hydrostatic head. This will depend upon the density of the liquid above, which will be about 1 gram per cubic centimeter.

In some cases, the liquid becomes trapped within the zone and cannot escape into the adjacent zones. Since the liquid has limited compressibility, the overburden will eventually become supported partially or wholly by the liquid and not by the rock matrix. As this occurs, the pressure of the liquid increases. It can become as high as the overburden pressure, when all of the overburden is supported by the liquid.

Normally, the density and acoustic logs will show a gradient which is a function of the overburden pressure. In an overpressured situation, however, the density and acoustic readings will depart from the normal gradient and remain nearly constant or very low. This departure on the log defines an overpressured zone, where the overburden is being supported by the relatively incompressible formation water. It is best seen in the shales. Logs showing overpressure effects are shown as Figure 4.11.

4.7 Saturation Evaluation

The formation factor, F_r, is related to the porosity:

$$F_r = a\,\phi^{-m}\left(\text{or, } F_r = \frac{1}{\phi^m}\right) \qquad (4.31a)$$

This is plotted in Figure 4.7. If the porosity and "a" are constant, then

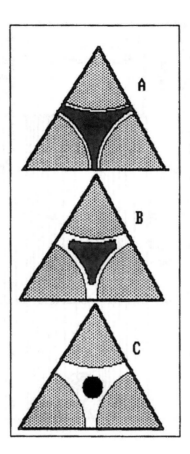

FIGURE 4.8
The several types of two-phase immiscible fluid saturation of the pore space. (Courtesy of SPWLA.)

$$R_o \propto R_w \qquad (4.37)$$

and, if the water resistivity, R_w is constant, then R_o is an inverse function of the porosity. Then, from Equations 4.29 and 4.30:

$$S_w = \left(\frac{R_t}{R_o}\right)^{-1/n} \qquad (4.38)$$

In a similar manner, if the residual oil saturation, S_{or} in the flushed zone is zero, the flushed zone water saturation, S_{xo} is

$$S_{xo} = \left(\frac{R_{xo}}{R_{mf}}\right)^{-1/n} \qquad (4.39)$$

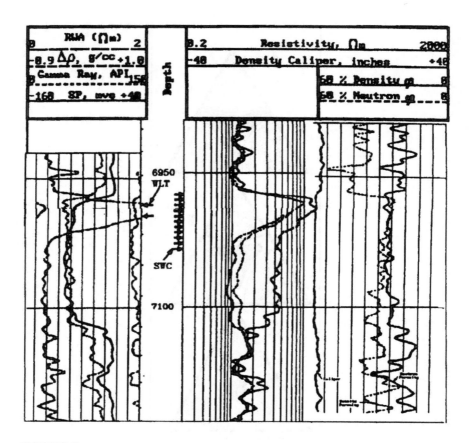

FIGURE 4.9
An electric/porosity log showing upward hydrocarbon migration.

and

$$S_{or} = 1 - S_{xo} \qquad (4.40)$$

where S_{or} is the relative amount of oil (or gas) left in the pore space when the free oil (or gas) is moved out of the zone by the invading mud filtrate). The amounts will depend upon the relative wetting abilities of the hydrocarbons and the water, and the capillary pressures. Values of S_{xo} are often estimated to be near 80%. Cutoff values of S_w are frequently estimated to be near 50%. These are good, rough approximations when we have no information about their actual values.

The difference between the oil saturation, S_o, and the residual oil saturation, S_{or}, is the moveable oil saturation, MOS, and is an indicator of the possible amount of producible oil:

$$MOS = S_o - S_{or} \qquad (4.41)$$

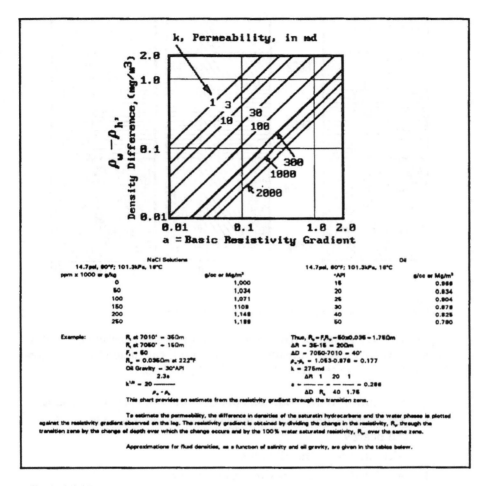

FIGURE 4.10
Estimating the resistivity gradient. (Courtesy of Schlumberger Well Services, Inc.)

4.8   Demonstration

If some of the formation parameter values are arbitrarily changed, we can observe the effect upon the other values. Equation 4.27 showed:

$$F_r = \frac{R_\varrho}{R_w} \tag{4.27}$$

If $R_o = 5$, $R_w = 0.2$, then $F_r = 25$.
If $R_o = 5$, $R_w = 0.5$, then $F_r = 10$.

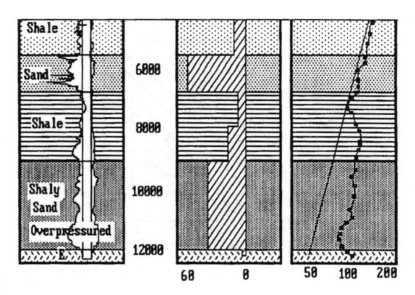

FIGURE 4.11
An illustration (hypothetical) of overpressure effect upon logs.

F_r is inversely proportional to the formation water resistivity, R_w. R_o is measured with the deep resistivity devices or it is calculated from a porosity value. There are several options for R_w. R_w will be:

1. Determined from a water sample from the zone you are working
2. Determined from a water sample from the same zone in another well nearby
3. Calculated from the SP value
4. Calculated from another resistivity curve
5. Determined from a crossplot

In an infinite volume of water, $F_r = 1.00$. That is its minimum value. In any other environment, F_r is greater than 1.00. F_r is always directly proportional to R_o. If the value of R_o is doubled, the value of F_r is doubled. F_r is also directly proportional to $1/Rw$. If R_w is doubled, then F_r is halved.

The calculated value of the porosity (the apparent porosity), ϕ_a, is a function of the value of m, the cementation exponent:

$$\phi_a = \left(F_r\right)^{-\frac{1}{m}} \tag{4.31b}$$

If $F_r = 25$, $m = 2.0$, then $\phi = 0.20$;
If $m = 1.8$, then $\phi = 0.17$;
If $m = 2.2$, then $\phi = 0.23$.

Combining Equations 4.27 and 4.31 shows that calculated value of the porosity, ϕ, is a function of the formation water resistivity, R_w

$$\phi = \left(\frac{R_w}{R_o}\right)^{\frac{1}{m}} \tag{4.42}$$

If R_o = 5.0 and m = 2.0, then ϕ = 0.20 when Rw = 0.20
If R_w = 0.50, then ϕ = 0.32.

Similarly, the value of the measured formation resistivity, when saturation is 100%, R_o, is an inverse function of the porosity, ϕ,

$$R_o = R_w\left(\frac{1}{\phi^m}\right) \tag{4.43}$$

If ϕ = 0.20, m = 2.0, and R_w = 0.20, then R_o = 5.0
If ϕ = 0.25, then R_o = 3.2
If m = 1.7 at the last porosity (ϕ = 0.25), then R_o = 2.1

From Equation 4.29,

$$I_r = \frac{R_t}{R_o} \tag{4.28}$$

we see that the ratio of the formation resistivity, R_t, to the formation resistivity at 100% saturation, R_o, is the resistivity index, I_r. When the zone is 100% water saturated, the value of R_t is, by definition, equal to R_o. Therefore, when the water saturation, S_w, is 100%, the value of I_r is minimum (1.00).

If R_t = 0.5, R_o = 0.5, then I_r = 1.00
If R_t = 10 and R_o = 0.5, then I_r = 20.0
If R_t = 100 and R_o = 0.5, then I_r = 200

From Equation 4.30, we see that the saturation, S_w, is an inverse function of the value of I_r and is a function of the saturation exponent, n:

$$S_w = \left(\frac{1}{I_r}\right)^{\frac{1}{n}} \tag{4.31b}$$

Therefore, if $n = 2$ and $I_r = 1.0$, then $S_w = 1.0$.

If $I_r = 20$, then $S_w = 0.22$
If $I_r = 4.0$, then $S_w = 0.50$
If $I_r = 2.0$, then $S_w = 0.71$
But, if $n = 1.9$ and $I_r = 2.0$, then $S_w = 0.69$

Similar relationships hold for the flushed zone. The value of R_{xo} replaces the value of R_t. The value of S_{xo} replaces that of S_w, and so forth. There is not an official special notation for the 100% water saturated value of the flushed zone resistivity, R_{xo}. In this text, however, the notation, R_{xos}, will sometimes be used for the value of R_{xo} when the zone is completely saturated with water (mud filtrate). This will correspond to the deep zone value of R_o.

4.9   Water Cut

The term "water cut" denotes the relative proportion of water which is produced along with the oil or gas from a well. The degree of water cut determines the viability of the well. The actual value or profitability of the well will depend upon the water cut and upon many other factors, both physical and economic. The amount of water produced will determine if the well can be produced or must be abandoned. Water is produced along with the hydrocarbon when production is from the zone of transition between the production of water-free hydrocarbon (at or near irreducible water saturation) and the lower water zone (at irreducible oil saturation).

The amount of water cut will depend upon the amount of water present which is not at the irreducible level. It will also depend upon the type of fluid which wets the rock of the matrix. The type of rock, the pore size, and the capillary pressure relationships all have an influence.

The relationships following, for clean and shaly sands, are from the *Dresser Atlas Chart Book*, 1985 (Atlas Wireline). They have also been used in nonfractured, nonvugular carbonates. They may help determine the degree of water cut.

4.9.1   Water-Cut Determinations

Whenever the economics of a particular well will allow or demand, zones in transition are sometimes perforated. Reference to a zone in transition pertains to that section of a hydrocarbon-bearing interval where production will be a combination of both hydrocarbons and water. A zone above transition is a zone at or near irreducible water saturation which, if formation conditions allow production, will produce water-free hydrocarbons. The amount of water produced by an interval in the transition zone determines whether the interval should be produced for profit or abandoned.

The following equations are applicable to both clean and shaly sands and have also been used in nonvugular, nonfractured carbonate rocks. The anticipated water cut (WC), for any well test, completion, or production attempt is given by:

$$WC_o = \frac{WOR}{1+WOR} \qquad (4.44a)$$

$$WC_g = \frac{WGR}{1+WGR} \qquad (4.44b)$$

where

$$WOR = (B_o)\frac{\mu_o k_w}{\mu_w k_o} \qquad (4.44c)$$

$$WGR = (B_g)\frac{\mu_g k_w}{\mu_w k_g} \qquad (4.44d)$$

where

WOR	= water to oil ratio
WGR	= water to gas ratio
B_o, B_g	= reservoir volume factor for oil and gas
k_w, k_o, k_g	= effective permeability to water, oil, and gas
μ_w, μ_o, μg	= viscosity at reservoir conditions for water, oil, and gas

Amyx, Bass, and Whiting define the *oil formation volume factor*, B_o, as the ratio of the volume of the oil at reservoir temperature, V_0, to the oil volume in the stock tank, V_{st}. The *gas formation volume factor*, B_g, is

$$B_g = \frac{ZRT}{V_m P} \qquad (4.44e)$$

where

Z	= the compressibility factor
R	= the universal gas constant
T	= the reservoir temperature in °F

P = the reservoir pressure in psia

V_m = the modal volume for the standard conditions desired

The ratios k_w/k_o and k_w/k_g can be related to the relative permeabilities to water, oil, and gas by the equations from *Dresser Atlas Log Interpretation Charts*, 1985 (originally from Jones, 1945 and from Boatman, 1961):

$$\frac{k_{rw}}{k_{ro}} = \frac{\dfrac{k_w}{k}}{\dfrac{k_o}{k}} = \frac{k_w}{k_o} \tag{4.45a}$$

$$\frac{k_{rw}}{k_{rg}} = \frac{k_w/k}{k_g/k} = \frac{k_w}{k_g} \tag{4.45b}$$

where

$$k_{rw} = \left[\frac{S_w = S_{wi}}{1 - S_{wi}}\right]^3 \tag{4.46}$$

$$k_{ro} = \left[\frac{0.9 - S_w}{0.9 - S_{wi}}\right]^2 \tag{4.47}$$

$$k_{rg} = \left[1 - \frac{S_w - S_{wi}}{1 - S_{wi}}\right]\left[1 - \frac{S_w - S_{wi}}{\left(1 - S_{wi}\right)^{1/4} S_w^{1/4}}\right]^{1/2} \tag{4.48}$$

where

k = absolute permeability

k_{rw}, k_{ro}, k_{rg} = relative permeability to water, oil, and gas

k_w, k_o, k_g = the absolute permeability to water, oil, and gas

5

Borehole, Mud, and Formation Effects

5.1 Borehole Effects

Thus far, we have covered some of the general factors that will be encountered with respect to the formations. We will now begin to examine the factors which will be encountered due to the presence of the borehole.

When a hole is drilled into the earth, the earth material removed as the hole progresses, is scraped, chiseled, and hammered out-of-place. Mechanical work is done on the rock material in the vicinity of the borehole. See Figure 5.1. This work can and often does damage the formation around the borehole. Drill cuttings are highly damaged by fracturing, washing, sorting, and in many other ways. The core may be damaged in similar ways. The formation material which we examine, when we analyze cores or make geophysical logs, has been altered by the drilling process and is not the same as the untouched formation which might contain hydrocarbon. Any of the samples from the earth will have been altered to some extent.

The borehole is filled with a drilling fluid which is continuously pumped into the hole from the surface, at near-surface temperature. It may not, at all, resemble the native fluids. This fluid is pumped down the drill pipe and up the annulus. (In a reverse circulation system, the circulation will be down the annulus and up the drill pipe; but the situation is the same.) The function of the drilling mud is to cool and lubricate the drill bit, carry cuttings out of the hole, and to counter the hydrostatic head of the formation.

Unfortunately, coring and logging needs are not the most important consideration. The mud is, however, the means of making electrical contact with the formation, and the borehole fluid characteristics influence *all* of the borehole geophysical measurements. The liquid portion of the mud, the mud filtrate, will invade the formation and wholly or partially displace the native fluids. This is shown schematically in Figure 5.2. The solid portion of the mud, the mudcake, will deposit on the wall of the hole at the permeable zones.

The drilling fluid or mud may be water, water/clay slurry (the most common type and the one which will be assumed in this text, unless otherwise stated), water phase emulsion, oil phase emulsion, oil, foam, air (or

FIGURE 5.1
A typical petroleum
drilling/coring setup.

some other gas), or some combination of these. It will contain some of the
fluids and solids of the zones through which the drill has already passed.
The mud pit or tank allows most, but not all of the solids to settle out of
the mud before it is pumped back into the hole. If a mud engineer is
present, he will monitor the quality of the mud as it comes out of the hole.
He will attempt to keep its condition constant or optimal. A mud logger is
not usually present, especially on non-petroleum holes. Very often, even
on non-petroleum holes, the cost of a good mud engineering service can
be recovered by the lack of hole problems.

Often, the mud will be oil or an oil-phase emulsion. These are used if the
water of the mud might damage the formation. Electric logs (except the
induction log) cannot be run in this type borehole environment. The other
log types and the other methods (cores, cuttings, etc.) may still be used.
Invasion is usually, but not always, shallow when these muds are used.
Special organic materials may have to be used on downhole equipment.

If the formation hydrostatic head is very low, or if the permeability is
very low, air (or other gas) or foam may be used to drill. Foam is often
used, because it can easily be circulated to carry out the cuttings. It does
not require the pressure and velocity for this purpose that gas does. The
lower velocities are easier to work with, also. Standard electric logs will
not operate in air filled holes. All other logging and coring devices may be
used in these fluids There will be little or no invaded zone in this type hole.

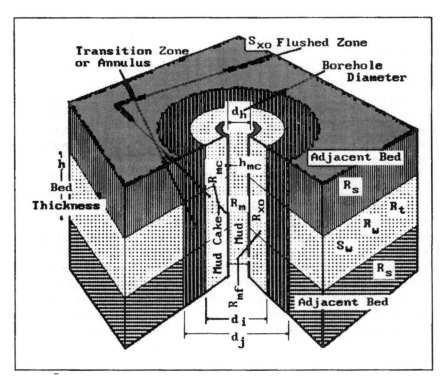

FIGURE 5.2
An idealized invasion diagram showing formation parameters. (Courtesy of Western Atlas Logging Services, Division of Western Atlas International, Inc.)

Frequently, the mud in the hole may be "topped off" or an air-filled hole filled with water just before logging. The water or mud should be treated the same as a mud, as far as logging or coring are concerned, except that the invasion is uncertain and may still be progressing rapidly during logging. Also, there may be sharp changes in borehole fluid salinity from one place in the mud column to another.

Occasionally brine is used in or as a mud. This is done when the formation clays are prone to swell; when they are partially desiccated. Mud resistivities will be very low (on the order of 0.02 to 0.05 ohmmeters). Mudcake is usually very thin and the invasion shallow in this type hole. The type of resistivity tool used in this environment will be different from that used in a less saline mud. Induction logs and the SP are useless in this type borehole environment. The focused resistivity systems are very good. The neutron systems are affected by the chlorine content of the mud, but are usually usable. The major salt in formation waters and in drilling fluids is sodium chloride, NaCl. KCl or other salts may be used in the mud and will cause high background gamma ray readings. Calcium compounds are frequently used in the mud in some areas. NaOH may be added as a preservative and can damage equipment.

To be sure that the formation fluids are contained within the formation, the density of the drilling mud is usually adjusted to make the hydrostatic head of the mud slightly higher than the formation hydrostatic head. Solids (clay) are added to increase the weight and viscosity of the mud. The higher viscosity will insure that the cuttings are carried out of the hole. Weighting materials, in addition to the clay, may be added. High viscosity will also help suspend weighting materials. It will, however, require more power from the drilling engines.

Any one of several additives may have been put into the mud to reduce or increase viscosity, reduce water loss, increase density, decrease foaming, increase foaming, prevent spoilage, reduce clay swelling, prevent flocculation, reduce lost circulation, and many other things. The driller's record will tell what has been done and why. It should be examined and become part of the record for the logs for that well. Some of these additives can affect the core and log results. This information, in abbreviated form, should be on the core record and the log heading. However, it seldom is.

The symbol for the resistivity of the mud is R_m and *must* include the temperature at which R_m was measured. The resistivity of the mud is usually measured and listed on petroleum electric logs. It almost never is on other types of logs. The resistivity of the liquid portion of the mud is the mud filtrate resistivity, R_{mf}. The mudcake resistivity symbol is R_{mc}.

5.2   Mud Logging

The examination and logging of the solid cuttings of the formation rock brought to the surface by the mud are usually done by the project geologist. This examination can furnish valuable information for formation evaluation and will be discussed further in Chapter 6. These cuttings are separated from the mud as the first step. The solid material not removed from the mud will settle out of the mud when the flow rate decreases in the mud pit.

Mud-logging service determines the constituents of the returning drilling fluid. Originally, this was done to monitor the mud quality. Since the drilling mud, however, picks up all of the material cut out of the formation by the drill, solid, liquid, and gas, measurements of these materials provides valuable information. Mud logging is mostly restricted to petroleum business. A sample mud log is shown in Figure 5.3. The purpose of mud logging is to monitor the condition or quality of the drilling mud. This allows proper correction of deviations from conditions which give efficient, safe, and rapid drilling. The mud is also monitored for evidence of the presence of hydrocarbons.

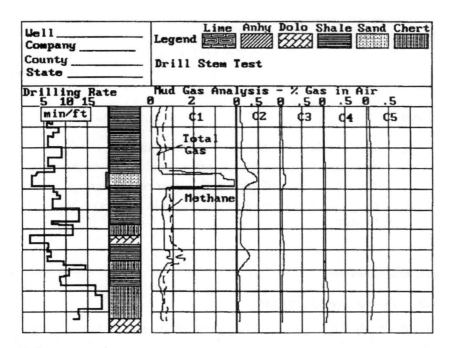

FIGURE 5.3
A typical mud log.

Besides containing formation rock material, the drilling mud will have become contaminated with the formation fluids through which the drill passes. The formation water in the mud is monitored because it will change the mud weight (density), viscosity, pH, and salinity. The electrical resistivity, R_m, which is a function of the salinity, is also useful in wireline logging. These values are measured by the mud logging service. Continuous recordings (logs) may be made.

The measured mud information may be presented in table form or it may be shown as a chart or a line graph log as a function of depth. A typical mud log may be presented with a stratigraphic column from the cuttings, the driller's log, an electric log, or one of many other options. See Figure 5.3.

5.2.1 Lag

Mud measurements must take into account the time necessary for the sample to reach the surface after a zone increment has been drilled. In the English system, the probable lag time, L, in minutes, is:

$$L = \frac{4.1 \times 10^{-2} D_h \left(d_h^2 - d_p^2\right)}{R_p} \tag{5.1}$$

or, in the metric system,

$$L = \frac{1.256 \times 10^{-5} D_h \left(d_h^2 - d_p^2 \right)}{R_p}$$
(5.2)

where

D_h = depth to the formation zone in feet or meters

d_h = diameter of the borehole or casing in inches or centimeters

d_p = diameter of the drill pipe in inches or centimeters

R_p = the pumping rate of the mud pump in gallons or liters per minute

5.3 Hydrocarbon Logging

Quantitatively logging the mud for hydrocarbons was first initiated in the United States in 1939. The format of the mud log is still not standard, although a standard format has been proposed by the American Petroleum Institute (A.P.I.). The mud log often contains a line graph of the liquid and gas hydrocarbon content and the content of other gases (i.e., CO_2, H_2, He, etc.). The presence of liquid hydrocarbons will usually be noted. The log will usually include comments (see Figure 5.3).

Gas logging systems usually consist of a sampler through which the mud is flowed. Here, the gas is extracted from the mud. The light fractions are extracted by agitating the mud. Small samples of mud are often taken and heated under controlled conditions to extract the heavier fractions. These methods are effective with most types of muds, including the oil base and oil phase emulsions. The amount of the extracted gas is often measured with a flowmeter. The gas is then routed to a detector to determine its composition. The detector may be one or more of several instruments.

A hot-wire Wheatstone's bridge may be used to detect the change in resistance of the heated leg of the bridge. The wire of that leg is chosen for a large temperature coefficient of resistance. Inert gases flowing across the hot wire will lower the temperature of the wire. Combustible gases will raise the temperature, due to their heat of oxidation. Figure 5.4 shows a diagram of a hot-wire Wheatstone's bridge.

Another method burns the gases in a small hydrogen flame. The conductivity of the burning gases is measured and interpreted as quantities of organic gases. This method has the advantage of being insensitive to noncombustible inorganic gases and carbon dioxide. The presence of methane

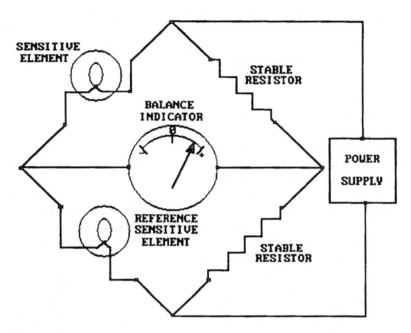

FIGURE 5.4
Schematic of a hot-wire Wheatstone's bridge.

may be determined with an infrared absorption meter. This method compares the absorption of a beam of infrared light through a cell with that through a reference cell. Only one type of gas at a time may be checked with this device.

A gas chromatometer may be used to separate the hydrocarbon constituents. The gases are mixed with a carrier gas and passed through a partition column. There they are separated by virtue of their different flow rates.

Liquid oils are usually detected in the mud by their fluorescence under ultraviolet light. The color of the fluorescence is an indication of the gravity (density) of the oil. This method is obscured by the usual low concentrations of natural oils in the mud and the contamination of drilling rig lubricants.

5.3.1 Sources of Gas

The gases detected in the mud have several possible sources. The heat and mechanical action of the drill bit and the catalytic action of the clay in the mud may produce gas from the formation and cuttings. Or, it may originate by thermal degradation of the mud components and/or minerals and other gases. If there is any gas in the formation, it will be produced if the hydrostatic head of the mud column is lower than that of the formation. Gas may be artificially or spuriously introduced into the mud. Spurious gas may appear from the formation due to the swabbing effect when the kelly is raised; it may also accumulate during any time that the mud pumps are shut down. Gas may also migrate from other zones or distant zones along faults and fractures.

5.3.2 Gas Deflection

The amplitude of the gas deflection on a mud log is a function of the amount of liberated gas per unit volume of mud and is presented as a function of hole depth. There may also be a background value due to the residual recirculated gas dissolved in the mud. Gas which was not removed from the mud on the first trip may be recirculated, from the mud-pit, down the hole. This will contribute to the background and may cause false secondary anomalies. A second, smaller deflection may appear one lag unit below the main deflection. This latter deflection may be due to the recirculated gas from the original zone.

During the drilling process, the drill bit introduces a volume of formation material into the mud. This volume is determined by the size of the drill bit and the thickness of the zone. Normally, the gas deflection is a function of the amount of gas present and the penetration rate. Since the log presentation is the amount of gas per unit volume of mud, a slower penetration rate will reduce the amplitude of the gas deflection for the same amount of gas present in the formation. See Figure 5.5. Crushing formation rock liberates contained hydrocarbons if they are present in the formation. These are directly liberated into the mud or they may be retained in the chips. If they are retained in the chips, some gas is liberated by expansion due to the decrease of pressure on the way to the surface. The resulting deflection amplitude on the gas log will be a function of amount of the liberated gas plus the background, minus the amount of the recirculated gas.

Partial liberation may occur if the gas is present only above the water in the zone. In this case, the gas deflection will be thinner than the zone thickness.

If the formation produces gas, it will be in addition to the liberated gas. The produced gas amounts will increase as the bit penetrates farther into the zone and may continue after the bit has passed through.

5.4   Analysis of the Drilling Data

The problem of drilling into an overpressured zone is a serious one. It can lead to costly action, delays, and potential blowouts. As we have seen, the overpressured zone can be detected from the wireline logs. The density of

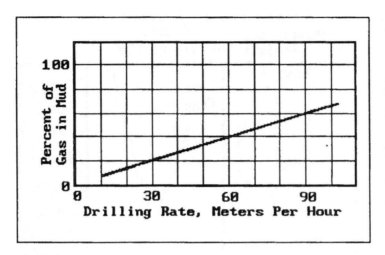

FIGURE 5.5
A typical response of amount of gas in the drilling mud, as a function of drilling rate.

the shale cuttings can be measured, and normally, the density of the shale will increase predictably with depth. Departures from the normal trend may indicate an overpressured zone. It can also be detected from the drilling data. One method is the *d*-exponent method:

$$d_{\text{English system}} = \frac{\log\left(\dfrac{R}{60N}\right)}{\log\left(\dfrac{12W}{10^6 D}\right)} \tag{5.3}$$

$$d_{\text{Metric system}} = \frac{\log\dfrac{R}{18.29N}}{\log\dfrac{10.41W}{D}} \tag{5.4}$$

where

R = the penetration rate in feet or meters per hour
N = the rotational speed (rpm) of the drill
W = the weight on the bit in pounds or kilograms
D = the bit size in inches or centimeters

A departure of the value of d from a uniform rate of change indicates a possible overpressured zone. A typical overpressure indication is shown in Figure 5.6.

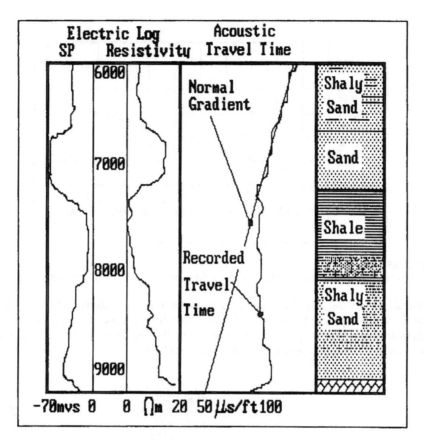

FIGURE 5.6
An example of logs in an overpressured zone (hypothetical).

A method in which density is calculated from the porosity, using an assumed rock matrix density, will show departures from the normal trends of values.

5.4.1   Shale Factor

Checking clay composition of the cuttings with methylene blue will give a shale factor. This quantitatively identifies the amount of montmorillinite in the shale and can be equated with the cation exchange capacity (CEC). The CEC value will be discussed in any text on the Spontaneous Potential.

5.5   Invasion

Once the drill penetrates a permeable zone, the normally higher borehole pressure, compared to the formation fluid pressure, will cause the mud to invade the formation. See Figure 5.2. Immediately, the solids in the mud (clay, silt, cuttings, weighting material, lost circulation material, etc.) will deposit on the wall of the hole and start to seal it off. This is the mudcake; its resistivity is R_{mc}. The low permeability of the mudcake will restrict the entry of the mud into the formation. Only the liquid portion of the mud (the mud filtrate), whose resistivity is R_{mf}, will continue, at a reduced rate, to invade the formation zone and partially or wholly displace the native fluids. The low and decreased permeability of the mudcake will rapidly restrict the flow of liquid into the formation. The flow will be quickly limited by the decreasing permeability of the mud cake and will soon drop to near zero.

The depth (horizontally) of the invasion will be almost independent of the permeability of the formation, but will be controlled by the permeability of the mud cake. It will depend upon the water loss of the mud and the porosity of the formation. The water loss is the rate of flow of the mud filtrate through the mudcake. Its units are in cubic centimeters per 30 minutes with a differential pressure of 100 psi. The higher the water loss, the greater the diameter of the invasion; the greater the porosity of the zone, the shallower the invasion.

The mud filtrate, whose resistivity is R_{mf}, (Figure 5.2) will enter the permeable zone and displace the native fluids immediately around the borehole. This is the "flushed zone" portion of the invaded zone. Its electrical resistivity is R_{xo}. If the flushed zone contains no hydrocarbon nor gas, it is saturated with mud filtrate. This special case will be labelled R_{xos} in this book. This is not an official designation, but is used to clarify explanations. The fluid saturation of the pore space, in this region, is S_{xo}. It, as with all saturation values, is a fraction of the pore volume. If hydrocarbon or gas was originally present along with the formation water, a small or residual portion, will be left behind. This is the residual oil or gas. The residual oil or gas saturation of the pore space is S_{or} or S_{gr}. The residual oil or gas saturation is typically less than 20% of the pore fluid.

Deeper into the formation, but still in the invaded zone, is the zone which is the interface between the formation fluid and the invading fluid. This is the mixed zone and its resistivity is R_i. This, as you might suspect, is a very difficult zone to work in. The actual extent of and distribution within this zone are difficult to determine. The spontaneous potential measurement is highly influenced by the total salinity ratio (R_{mf}/R_w) across this interface. We will frequently use the SP to determine the value of R_w, the resistivity of the formation water.

Beyond the mixed zone is the deep, undisturbed zone. This zone is of great enough extent to be a "sink" and can absorb the relatively small disturbances caused by the borehole, without change. This is where we expect to find the bulk of the oil or gas. Its fluid is the formation water and any hydrocarbon or gas. The resistivity of the formation water is R_w. The corrected, measured resistivity of this zone is R_t. A special case of R_t, when the zone contains no hydrocarbon nor gas is labelled R_o. R_o is the corrected, measured resistivity in this zone when $S_w = 1.00$. The formation water saturation of the pore space in this zone is S_w. Any oil or gas which might be in the pore space will have a saturation of S_o or S_g.

We will be able to get a mud sample on the surface and measure its resistivity, R_m, for use with the resistivity and SP curves. The values of R_{mf} and R_{mc} are also needed for the same purposes. The most accurate way to determine these values is to separate the mud components with a mud press and measure the values with a resistivity meter. Figure 5.7 shows a Fann mud press.

FIGURE 5.7A
A standard Fann mud press. (Courtesy of Fann Instrument Company.)

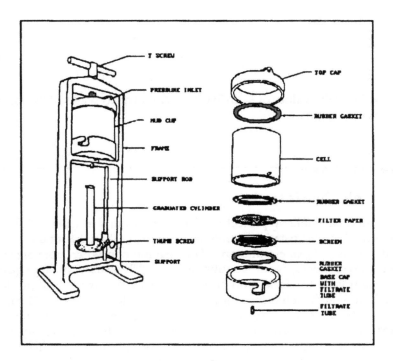

FIGURE 5.7B
An exploded view of the Fann Mud Press. (Courtesy of Fann Instrument Company.)

TABLE 5.1

Values of K_m, the Mud-Weight Operator

Mud (lbs/gal)	Weight (kg/m³)	K_m
10	1200	0.847
11	1320	0.708
12	1440	0.584
13	1560	0.488
14	1680	0.412
16	1920	0.380
18	2160	0.350

Courtesy of Schlumberger Well Services, Inc.

It is not always possible to use a mud press, however. The values may be estimated, using one of several available methods. One of the better methods is the SWC relationship (see Table 5.1),

$$R_{mf} = K_m \left(R_m \right)^{1.07} \qquad (5.5)$$

and

$$R_{mc} = 0.69 R_{mf} \left(\frac{R_m}{R_{mf}} \right)^{2.65} \tag{5.6}$$

You will find the values of K_m in Table 5.1 and in chart books of the logging service contractors.

5.6   Hole Considerations

If the mud cake fails to protect the formation from the drilling fluid and the formation hydrostatic pressure is much lower than that of the drilling mud, we have "lost circulation". This can happen when pore sizes are large, as with fractures and vugular porosity. Whole mud, in large quantities, can invade the formation in this situation. The driller will put large area, light-weight solids into the mud, to try to block the holes. The material may be cottonseed hulls, nut hulls, paper, rags, straw, rice hulls, or anything he thinks will stop the lost circulation. This is necessary, but it makes logging and coring very difficult. It poses the danger of sticking and losing logging tools and/or drilling equipment.

An ever-present danger, particularly when the borehole is not straight, is that the drill pipe, the logging tool, or the logging cable will cut through the mud cake. The caliper tool (Figure 5.8) can be used to determine the quality of the borehole and the borehole wall. The pipe or tool can then be pressed to the wall of the hole with a large force. This force is proportional to the differential pressure across the mud cake and the area of the tool or pipe against the hole wall. If this happens during drilling, the drill pipe can easily be twisted off. Or, if the drill pipe sticks and then suddenly releases, it can unscrew a drill collar downhole. This also can cause severe problems recovering logging tools or drill pipe. For example, if a logging tool is 2 inches diameter and 10 feet long and the differential pressure is 100 psi, the force holding the tool to the wall of the hole can be as much as 12 tons.

Almost all geophysical logging devices are affected by the presence of the borehole, itself. A major part of our data reduction task will be to make corrections to the signal received at the surface for the presence of the borehole. These problems are caused by the inclusion of the signal components due to the size of the borehole, the characteristics of the fluid filling the hole, the quality of the hole (the roughness of the wall of the hole or rugosity, caving, bridges, etc.), the presence and characteristics of the mud cake. These borehole materials contribute signal components which generally are not of interest.

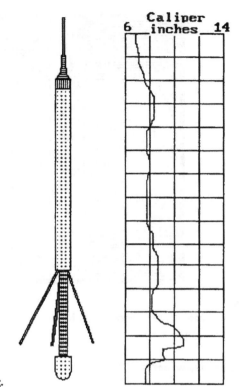

FIGURE 5.8
A three-arm caliper and log.

Many of the procedures in log analysis require good hole diameter and quality information. Many logging tools incorporate a simultaneous caliper, the purpose of which is to report on the diameter and quality of the hole. See Figure 5.8. With it one can determine the mud cake thickness, hole diameter, the locations of rugosity, caving, bridging, and ledges. This is especially important when working with sidewall tools. Tool response of other than sidewall tools, however, may also be affected by hole wall conditions. This is quite noticeable with some of the acoustic systems.

The measured volumes of all logging tools will include some portion of the borehole. See Figure 5.9. Thus, they are sensitive to the borehole diameter, d_h, and the borehole fluid characteristics. Even the sidewall tools are influenced. These latter are sensitive to the presence of mud cake. Since the mud and the mud cake are not ordinarily interesting to our project, except to require a correction, the response due to them must be removed from the signal or the data. If the signal is corrected for some of the borehole effects through the design of the logging tool or circuitry, the tool is said to be compensated. A major element in instrument design is to minimize the contribution of these unwanted components. This problem is also a major one during data reduction.

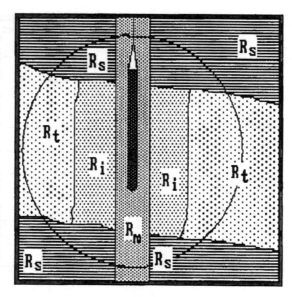

FIGURE 5.9
The volume encompassed by resistivity logging.

5.7 Logging-Tool Position

If a tool is allowed to hang free in the hole, it will normally lie against the wall of the hole. It is said to be eccentered (ek-sen'-terd). It may also be deliberately eccentered. In this case, there will be eccentering devices (bow springs, fingers, arms, or some device) to hold the tool against the wall. Some tools are deliberately centered. This is usually done with arms or bow springs. Some tools will have devices (stand offs) to hold the tool a few inches or centimeters away from the wall. See Figure 5.10.

The position of the tool in the borehole will affect the response. Charts are available from the contractor for centered and for eccentered situations. Consider the example of a natural gamma ray tool which has a 98% sphere of investigation in a sand of about 3 feet (1 meter). See Figure 5.11. In the centered gamma ray probe case, if the tool is centered in the borehole and the borehole is the same diameter as the probe, a correction factor will be 1.0, because there is no hole to correct for. As the hole diameter increases, the correction factor will become greater than 1.0 because there will be borehole fluid which will contribute only a small amount of radiation to the formation signal. It is also an attenuating medium between the detector and the formation. As the hole diameter increases, a smaller portion of the signal comes from the formation and more from the borehole. Therefore, the correction factor becomes larger. Finally, the hole will become so large that an insignificant part of the signal comes from the formation. At this

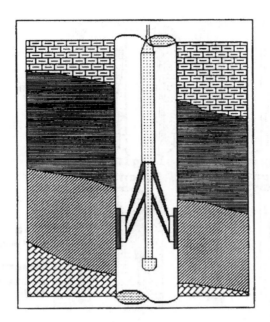

FIGURE 5.10
A typical sidewall probe.

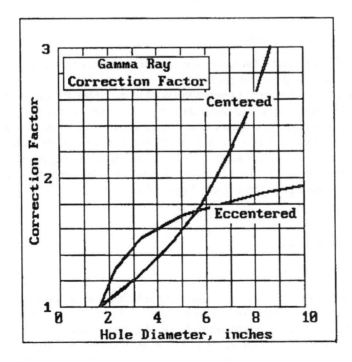

FIGURE 5.11
A hypothetical comparison of centered and eccentered GR borehole correction curves.

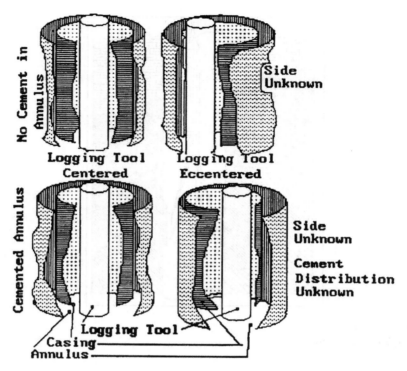

FIGURE 5.12
Cased-hole correction problems.

point, the correction factor is approaching infinity. Nearly all of the response is from the borehole fluid.

If the tool is eccentered (against the wall of the hole), the case will be the same when the borehole is the same size as the tool. As the hole diameter increases, the correction factor will again increase and be greater than 1.0. When the borehole has an infinite diameter, the wall of the hole is a flat plane, and the tool lies against it. Only about half of the signal comes from the borehole fluid and half from the formation. The correction factor, in this case is asymptotic to slightly higher than 2.

5.8   Cased Holes

The situation becomes more complex if the hole is cased. Many of the tools we use can and will be used in cased holes. It is not difficult to determine the change of tool response when the tool is run in casing. The tool position, however, is in doubt. A correction must be made and the data can be used. The reliability of the information, however, has been degraded by the positional uncertainties. This must be kept in mind while working with the data. See Figure 5.12.

If the casing is centered and the tool eccentered, then the tool will always have a standoff, from the wall of the hole, which is equal in thickness to the thickness of the annulus between the casing and the hole wall. The material of the annulus may be mud or cement. Occasionally, the material will be the centralizers or joints, but these are usually easy to identify.

If the casing is not centered, then it may be anywhere in the hole. If the casing rests on the bottom, it will have a corkscrew shape in the hole. The tool may be anywhere from eccentered to centered and its relative position will change constantly while logging.

Occasionally you will have a log made in a multiple-completion hole. In this case, there will be several strings of tubing in the hole filled with mud, gas, water, or cement. The log will have been made from one of the strings. The tool may be anywhere in the hole and it may have interference from the other strings of tubing.

6

Cores, Core Analysis and Cuttings

6.1 Introduction

Coring and core analysis constitute one of the oldest methods for gathering subsurface data. Figure 6.1 shows a "portable" diamond drilling machine. (Baroid, 1979). Diamond drill bits frequently cut a core along with routine hard rock drilling. Information from core samples is essential to the proper evaluation of both petroleum and non-petroleum formations. The analysis of core samples tells much which cannot be determined in other ways. As important is their function as the calibration standard for other methods, such as wireline geophysical logs. Correlation with other methods is important, as is combination of core data with that from other methods.

When Schlumberger first introduced commercial wire-line geophysical logging, it was called "Electrical Coring". Figure 6.2 shows a representation of the title page from a paper by Conrad and Marcel Schlumberger, describing "Electrical Coring" (Allaud and Martin, 1977). At that time it was presumed that the new method would eventually eliminate the need for coring. Today, coring-related services are more active, accurate, and widely used than ever before. Obviously, they have not been supplanted. Coring methods complement the other methods. All of the information is needed. No one method should be used without reference to similar data from other sources.

6.2 Uses of Coring

Table 6.1 shows some of the uses of coring (LeRoy and LeRoy, 1977). Two of the more important uses are the determination and verification of permeability and porosity. Borehole coring is done in many ways and for many reasons. Specialized equipment and techniques exist and new ones are devised frequently. Standard techniques and equipment range from the bailers used in cable tool holes to the sophisticated oriented coring devices used in scientific work.

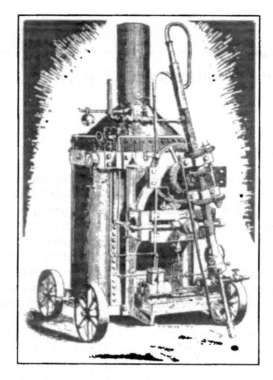

FIGURE 6.1
A portable diamond drill of the 1870 period. (Courtesy of Baroid Drilling Fluids, Inc.)

Conventional rotary coring uses a fairly standard drilling assembly at the bottom of the drill pipe. The cutting bit is a torus to cut away an annular ring. Figure 6.3 is a sketch of a diamond coring bit. The cutting surface may be imbedded diamond, a roller assembly, or a drag-type bit, depending upon the competency and hardness of the formation material and the job to be done.

As the cutting proceeds, the uncut center of the rock, or core, is pushed up into the coring assembly, through the open center of the bit. The core is pushed into a floating barrel in the center of the assembly; a "catcher" (usually steel fingers) keeps the core in the barrel. When the proper length of core is cut (determined by the capacity of the barrel and the length of pipe used), the assembly and core are retrieved by "tripping" the pipe (pulling the drill pipe up, removing the core, servicing the bit/core assembly, and returning it to the bottom). Core sizes vary from less than an inch (2.0 cm) to 6 inches (15 cm) in diameter. Lengths may be as little as 1 inch (2.5 cm) to more than 30 feet (9 m).

There are, of course, many modifications of this system to perform routine and special jobs. The problem of keeping all of the unconsolidated material is especially difficult. Split barrels and rubber or plastic sleeves are

THE ELECTRICAL CORING

PAPER SUBMITTED BY

C. et M SCHLUMBERGER

AT THE SECOND INTERNATIONAL DRILLING CONGRESS

Paris (September 1929)

SOCIETY DE PROSPECTION ELETRIQUE

(PROCEDES SCHLUMBERGER)

30. RUE FABERT - PARIS - VIIe

AND

SCHLUMBERGER ELECTRICAL PROSPECTING METHODS

25, BROADWAY - NEW YORK CITY

FIGURE 6.2
A representation of the title page of the 1st article on electrical logging.

sometimes used to help with this. One modification of the standard assembly allows the barrel to be retrieved and replaced with a wire line through the drill pipe. During the Mohole Project, a collapsible coring bit was designed which could be retrieved and replaced, along with the core barrel, using the wire line.

Sidewall percussion coring fires hollow bullets horizontally into the wall of the drill hole with explosive charges. They are retrieved by short wire cables when the gun is moved upward. The bullets, each with its small

TABLE 6.1

Core Analysis Data and Uses

Routine Data	
Data	**Use**
1. Porosity	Define storage capacity
2. Horizontal permeability	Define flow capacity, permeability distribution and profile
3. Saturations	Define (1) Presence of hydrocarbons (net pay and contacts) (2) Type of hydrocarbon (gas or oil) (3) Connate water if oil base mud is used
4. Lithology	Define rock type and characteristics of core (fractures, vugular, laminated, etc.)

Supplementary Tests

1. Vertical permeability	Define coning probability and gravity drainage potential
2. Core-Gamma log	Define lost core and depth relation of core with downhole logs (requires downhole gamma ray log)
3. Grain density	Define density log calculations
4. Water chloride	Define connate water salinity in oil base mud cores and degree of flushing in water base mud cores
5. Oil gravity	Estimate reservoir gravity from correlations based on retort oil gravity

Interpretations from Core Analysis

1. Prediction of fluid production (gas, condensate, oil, or water)
2. Definition of oil-water, gas-oil, or gas-water contacts and transition zones
3. Possibility of gas or water coning
4. Completion intervals

Courtesy Core Laboratories, Inc.

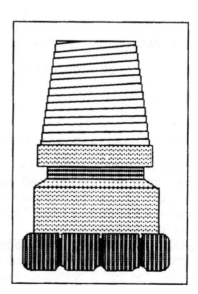

FIGURE 6.3
A sketch of a diamond core bit.

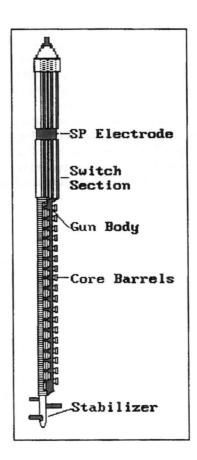

SP Electrode

Switch
Section

Gun Body

Core Barrels

Stabilizer

FIGURE 6.4
A percussion-type sidewall sample gun.

sample, drop into and hang in a cavity in the gun, until the gun is returned to the surface. Figure 6.4 is a sketch of a sample gun. The samples are small and likely to be damaged, but they are better than guesswork, if no conventional cores are available.

Hard-rock sidewall coring uses a motor and/or a hydraulic system to position and rotate a small coring bit into the wall of the hole. Once the core is cut, the bit retracts, breaking off the core. The core is then dropped into a magazine in the tool. Often, a ball is dropped on top of the sample in the magazine, to separate it from the next sample. One version of the hard-rock tool uses two diamond-faced circular saws to cut a V-shaped wedge from the wall. A system which is run on drill pipe, uses a gouge to pull a V-shaped strip from the wall in soft-formation materials. Obviously, mudcake is part of the sample and the fluid is mostly mud filtrate.

Formation-fluid samples may be taken in several ways. The driller can often obtain a sample of the formation water by placing packers in the well above and, sometimes below the permeable zone and flowing or pumping the fluid from the formation. The first fluid is from the invaded zone; later flow is from the undisturbed zone. This method is common in water-well drilling. A second fluid-sampling method is the production-type drill stem tests. These are done in a manner similar to the water sampling described previously. The goal here, however, is to sample the hydrocarbon content. Careful records are kept of the fluid amounts and types, as well as continuous records of the pressures.

A third method is the wireline fluid tester or formation tester. This is a sampling tool operated on the logging wire line. The fluid sample consists mostly of fluid from the invaded zone. It operates in moderate- to high-porosity zones. Pressure records are kept during the fluid flow. Figure 6.5 shows a drawing of a typical fluid sampler.

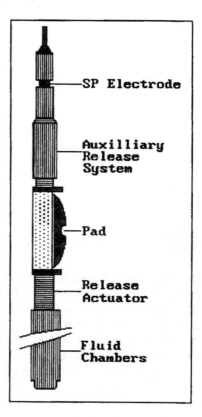

FIGURE 6.5
A sketch of a fluid sample taker.

In oceanography and in well drilling, a drop corer is sometimes used. Ocean-floor material or borehole bottom material can be sampled in this way. The corer is an open-ended (on the bottom) cylinder with stabilizing fins on the top to make it drop open end down. Inertia forces the corer into the bottom material. The device is retrieved with an attached wire line. In wells, the method is used mostly in cable tool holes.

6.3 Core Quality

Cores are severely flushed by the drilling mud before, during, and after the drilling operation. This alters the core fluids to an unknown extent. The type and amount of change of the core will also depend upon the type of drilling fluid. It is necessary to take this into account when using core derived values, such as the residual oil saturation, S_{or}. The effect of various drilling fluids upon the saturation of the core is shown in Table 6.2 (LeRoy and LeRoy, 1977).

TABLE 6.2

Effect of Coring Fluid on Core Saturations

Coring Fluids	Filtrate	Water	Hydrocarbons
Water base	Water	Increased	Decreased
Oil base	Oil	No change[a]	Replaced[b]
Inverted oil emulsion	Oil	No change[a,c]	Replaced[b]
Oil emulsion	Water	Increased[d]	Decreased
Hydrocarbon gas	Gas	No change[a]	Replaced
Air	Uncertain	Uncertain[e]	Decreased

[a] If sample is at irreduceable (immobile) water saturation prior to coring. Otherwise filtrate will displace water and reduce its saturation if in a transition zone. Precautions to remove all extraneous water from the system should be taken.

[b] This replacement sometimes causes gas sands to exhibit oil saturations at the surface indicative of oil rather than gas productive reservoirs.

[c] Inverted oil emulsion muds contain water. Loss of whole mud to high permeability formations will increase water saturations. Practical permeability limit for whole mud may be several hundred millidarcys.

[d] Improper mud control may cause filtrate loss to be oil rather than water.

[e] Data indicate that saturations may increase, decrease or remain unchanged, depending upon heat generated during coring, hole condition, and the agent mixed with the air.

Courtesy Core Laboratories.

Cooling of the core, as it is brought to the surface, will cause thermal contraction of the fluids in the pore spaces of the core. This will change the saturation values and will have a tendency to pull in borehole fluids. Figure 6.6 shows some typical changes in fluid saturation of the core during the trip to the surface (LeRoy and LeRoy, 1977).

Decreasing pressure during the trip to the surface will allow the contained gases to expand and some liquid-phase gases to boil. Some portion of them will be expelled. Even the pore volume can change with the decrease of overburden pressure. Figure 6.7 shows the effect of the removal of the overburden pressure on several rock types (LeRoy and LeRoy, 1977).

Soluble materials, such as halite, will be partially or wholly dissolved before, during, and after the cutting process. This can be slowed or eliminated by saturating the drilling mud with the material to be retrieved. Soft

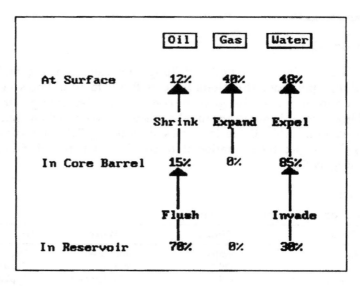

FIGURE 6.6
Typical core saturations changes during recovery. (Courtesy of Core Laboratories, Inc.)

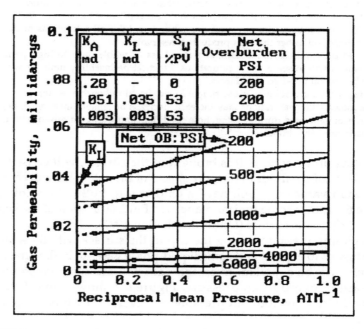

FIGURE 6.7
Permeability reduction due to gas slippage, irreducible water, and overburden pressure. (Courtesy of Core Laboratories, Inc.)

clays and young, shallow shales are easily lost during the coring operation. Some forms of coal are quite difficult to core because the drilling fluid easily breaks them up. Air may sometimes be used successfully as a drilling fluid in these difficult cases.

The mechanical forces used to drill the core can produce microfractures which will interfere with mechanical testing and porosity and permeability determinations. Myung (Myung and Helander, 1972) has shown that, for this reason, the mechanical parameters measured on core samples have a wider variance than do those determined from wire line logs.

High-quality cores can be severely degraded when they are improperly handled on the surface. Poor indexing, cataloging, labelling, and boxing procedures are a major cause of poor information. Sedimentary cores should be kept from drying. They should be enclosed in plastic bags or sleeves to slow drying, oxidation, and other forms of alteration. Sometimes freezing may help preserve the core. Photographing and running a core gamma log can help prevent and correct cataloging and correlation errors.

6.4   Core Analysis

Porosity, permeability, rock type, and other parameters may be accurately and positively measured on core samples. These are laboratory measurements, conducted in controlled environments and are often the standard against which other methods are compared. When carefully performed, they can be very precise and accurate. They can be indispensable. These tests are performed, however, on a small (1" × 1.5"; 2.5 cm × 3.8 cm) plug cut from the center of the core. The statistical variation of the resulting values will therefore be higher that those from other methods whose sample sizes are larger.

Determination of rock type from core samples is often only fairly good because of the small sample size, unless the measurements or determinations are made on the full size core. Magnetic, acoustic, and resistivity measurements are usually quite good.

6.5   Core Information

Preliminary information from the core analysis can be used statistically to predict conditions to be encountered. This can also often allow a coring program to be shortened. Equally important, it can sometimes indicate that additional coring, sampling, and/or wireline logging needs to be done.

The mean permeability and variations for the zone of interest can be plotted, well by well. This will quickly tell the range of permeability to be expected and whether the project is likely to be a successful one or whether special, additional treatment will be needed. Porosities can be used in the same way. Mean values can be established early and used for project wide analysis purposes. Sand or ore body thicknesses can be predicted to any degree of confidence, depending upon goals and amounts of information. Errors of measurement can often be detected and corrected. Calibration errors and information variances are easily normalized.

6.6 Laboratory Measurements on Cores

6.6.1 Extractors

A core sample must be cleaned of all contained liquids before porosity and permeability determinations can be made. This is commonly done in a modified ASTM extractor. This extractor collects the residual pore water by refluxing hot solvent vapors across the core, condensing, and collecting the water for volumetric and composition measurements. Figure 6.8 is a sketch showing the principle of this extractor. Drying is usually done by heating the core to 240°F (115°C). Hydratable clays are dried at low temperature and in a controlled humidity.

When all of the water has been removed, the sample may be further cleaned. Solvents used for cleaning must be chosen so as to not react with the rock of the sample. Solvents used are tolulene, benzine, and other organic solvents. Water or methanol is used to remove residual salt. Some of the cleaning methods are

1. A Dean Stark or Soxhlet extractor. This operates on the same principle except there is no provision to collect and measure the water. Figure 6.9 is a sketch of this extractor. This method is slow, but gentle.

2. Flushing in a centrifuge. This is fast, but requires a rugged sample.

3. Flushing with carbon dioxide and tolulene; a fast method.

4. Pressure flushing with a solvent. This method is slow.

5. Vapor soaking with tolulene.

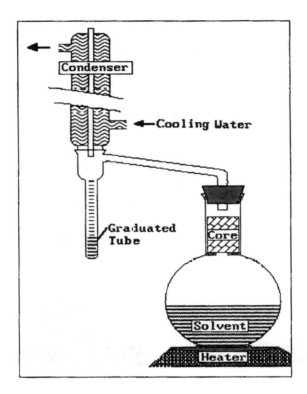

FIGURE 6.8
A modified ASTM extractor.

The collected water amount and the dry core can be used to determine the residual oil saturation, S_{or}. The saturation is given by:

$$S_{or} = \frac{W_{cw} - W_{cd} - W_w}{\phi \rho_o} \qquad (6.1)$$

where

W_{cw} = the weight of the wet core
W_{cd} = the weight of the dry core
W_w = the weight of the core water collected
ϕ = the fractional porosity
ρ_o = the density of the oil

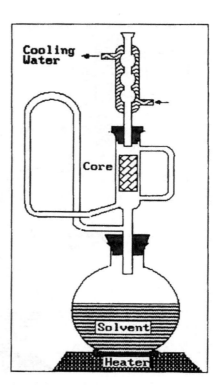

FIGURE 6.9
A Soxhlet extractor.

6.6.2   Retorting

A simple retort apparatus may be used to collect the core fluids, especially the heavy fractions of the hydrocarbons. This is a heated chamber in which the sample is placed. The vapors of the (normally) liquids condense and are collected and measured volumetrically.

Because of the high temperatures needed (up to 1200°F, 650°C), waters of crystallization and hydration may be driven off. Montmorillinite is particularly susceptible to dehydration. Solid organic materials may also be broken down. Both of these events may cause high readings and change the character of the rock matrix.

Retort results are,

$$S_{wr} = \frac{\phi}{V_w} \tag{6.2}$$

$$S_{or} = \frac{V_o}{\phi} \tag{6.3}$$

$$S_{gr} = 1 - S_{wr} - S_{or} \tag{6.4}$$

6.6.3 Bulk Volume

The bulk volume of a geometrically regular core, if it has a convenient shape, is simply calculated by measuring the sample dimensions with a caliper.

If the sample does not have a convenient shape, it can be weighed in air, then immersed in a non-wetting liquid, and weighed in the liquid. The difference in weights will allow the volume to be calculated.

6.6.4 Pore Volume — Boyles' Law Method

The volume of the effective pore space of the core may be measured by employing Boyle's Law. A container is filled with helium gas to a known pressure and temperature. The amount of gas required to fill the container describes its volume. The clean, dry sample is then placed in the container and the volume at the same temperature and pressure is again measured. The difference between the two measurements is the volume of the rock matrix material of the sample. The effective porosity of the core is:

$$\phi = V_{bulk} - V_{gas} \tag{6.5}$$

6.6.5 Pore Volume — Washburn-Bunting Method

The Washburn-Bunting method pulls air from the clean, dry sample with mercury and, at the same time, measures the volume of the air:

$$\phi = \frac{V_{atm} - V_{corr}}{V_{bulk}} \tag{6.6}$$

where

V_{atm} = the volume of extracted gas at atmospheric pressure

V_{corr} = the volume correction for gas adhering to the inside of the vessel

V_{bulk} = the bulk volume of the sample

6.6.6 Grain Density

When the grain density method is used, the clean, dry sample is coated with a thin, waterproof coating and is weighed. The bulk volume is then measured in a pycnometer, by weighing the displaced water. The bulk density is calculated. The coating is then removed, the sample crushed, and the volume of the grains is determined in the pycnometer. The grain density is calculated. Then:

$$\phi = \frac{\rho_{gr} - \rho_{bulk}}{\rho_{bulk}} \tag{6.7}$$

where ρ_{gr} is the grain density ρ_{bulk} is the bulk density of the sample.

6.6.7 Usual Practice, Pore Volume

Usual practice is to dry the core sample, weigh it the first time (1), saturate it with a liquid of known density (water, naphtha, etc.) and weigh it a second time (2):

$$\phi = \frac{M_1 - M_2}{\rho_{bulk} V_t} \tag{6.8}$$

where

M = the mass of the sample
ρ_{bulk} = the bulk density of that material
V_t = the total volume

The total volume is often measured by immersing the core sample in mercury. The surface tension of mercury will prevent it from entering the pores. Thus, the displacement is a measure of the bulk volume.

6.7   Core-Derived Permeability

Core permeability is measured by forcing gas or liquid through a contained sample. The permeability, in darcys is

$$k_a = -\frac{1}{\mu} \frac{dP}{dL} \tag{6.9}$$

where μ is the fluid viscosity, dP/dL is the change in pressure per unit length. This absolute permeability, k_a, assumes that the sample is saturated with a single fluid.

When there is a mixture of liquids or liquids and gases, as we have in petroleum reservoirs, the relative permeability of each component may be different. The relative permeability, K, is the ratio of the permeability of the sample of any specific fluid, k_n, to the absolute permeability, k_a:

$$K = \frac{k_n}{k_a} \qquad (6.10)$$

The relative permeability will determine the viability of the zone to be produced. Reservoir permeability is influenced by the overburden pressure. *In situ* permeability values have been reported lower than laboratory measurements by as little as 7% to as much as 100% (McLatchie and Hemstock 1958; Vairogs 1970).

6.7.1 Laboratory Measurements

Laboratory permeability measurements are made with the flow of a gas. The permeability to a gas, however, will vary with the pressure of the gas. Figure 6.10 shows this effect, as determined by Klinkenberg 1941 (Core Laboratories). Table 6.3 lists some of the Klinkenberg corrections.

Darcy's Law states

$$q = \frac{k_a A}{\mu L} \Delta P \qquad (6.10)$$

where

q = the rate of flow of the test fluid in cubic centimeters per second

A = the cross sectional area to the flow in cubic centimeters

μ = the viscosity of the fluid in centipoise

ΔP = the differential pressure in atmospheres across the test volume

L = the test length of the sample in centimeters

k_a = the permeability of the sample in darcys

If air is used as the test fluid, the compressibility must be accounted for:

$$k_a = \frac{q_m \mu L}{A(P_2 - P_1)} \qquad (6.11)$$

where q_m = cubic centimeters per second at average pressure. Usually, a core plug, one inch diameter, is used as a sample. Larger samples will use radial flow from the center of the sample:

$$k_a = \frac{q \mu \ln\left(\frac{r_e}{r_w}\right)}{2\pi h(P_e - P_w)} \qquad (6.12)$$

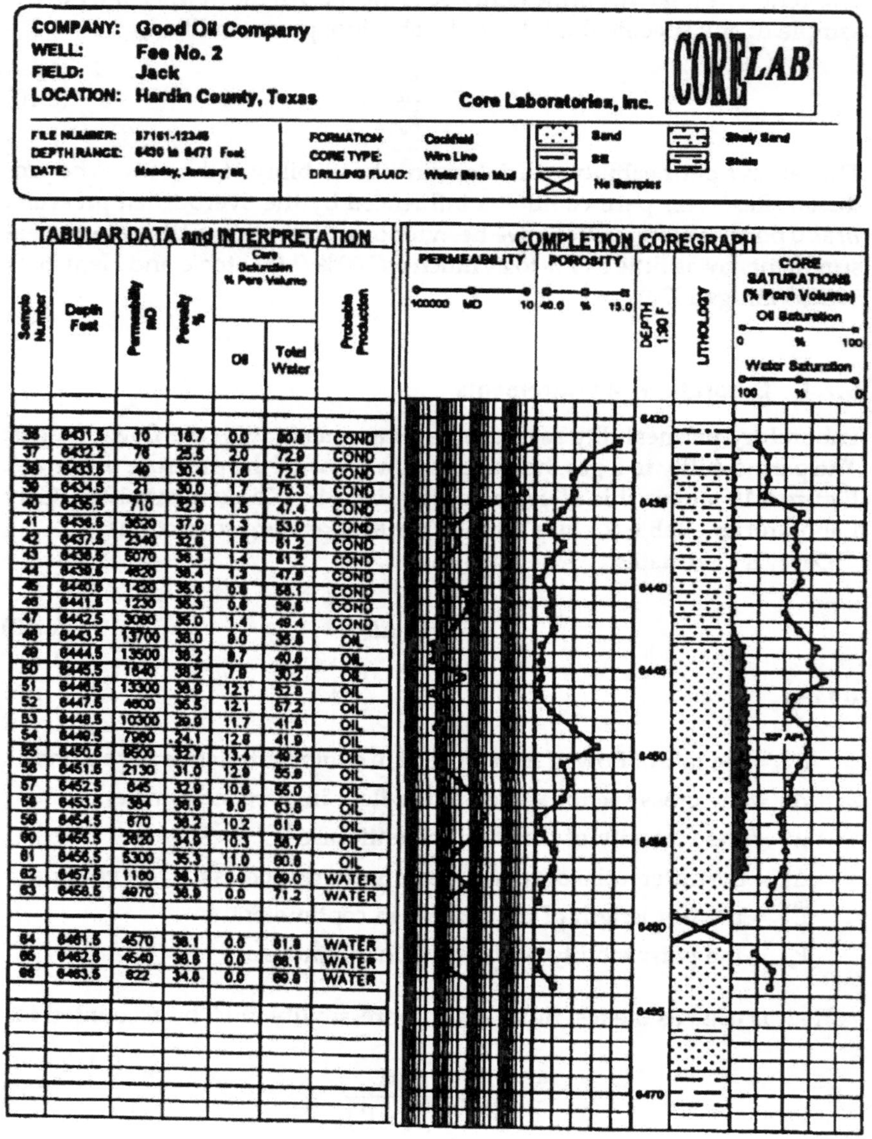

COMPANY: Good Oil Company
WELL: Fee No. 2
FIELD: Jack
LOCATION: Hardin County, Texas

Core Laboratories, Inc.

CORELAB

FILE NUMBER:	37161-12345	FORMATION:	Cockfield
DEPTH RANGE:	6430 to 6471 Feet	CORE TYPE:	Wire Line
DATE:	Monday, January 88,	DRILLING FLUID:	Water Base Mud

Sand · Shaly Sand · Shale · No Samples

TABULAR DATA and INTERPRETATION

Sample Number	Depth Feet	Permeability md	Porosity %	Core Saturation % Pore Volume — Oil	Core Saturation % Pore Volume — Total Water	Probable Production
36	6431.5	10	16.7	0.0	80.8	COND
37	6432.2	76	25.5	2.0	72.9	COND
38	6433.5	40	30.4	1.8	72.8	COND
39	6434.5	21	30.0	1.7	75.3	COND
40	6435.5	710	32.5	1.5	47.4	COND
41	6436.5	3820	37.0	1.3	53.0	COND
42	6437.5	2340	32.8	1.5	51.2	COND
43	6438.5	5070	36.3	1.4	51.2	COND
44	6439.5	4820	36.4	1.3	47.8	COND
45	6440.5	1220	36.8	0.8	58.1	COND
46	6441.5	1230	36.5	0.8	50.8	COND
47	6442.5	3080	36.0	1.4	48.4	COND
48	6443.5	13700	38.0	9.0	35.8	OIL
49	6444.5	13500	38.2	8.7	40.8	OIL
50	6445.5	1640	38.2	7.8	30.2	OIL
51	6446.5	13300	38.0	12.1	57.8	OIL
52	6447.5	4600	36.5	12.1	57.2	OIL
53	6448.5	10300	29.9	11.7	41.8	OIL
54	6449.5	7960	24.1	12.8	41.9	OIL
55	6450.5	9500	32.7	13.4	48.2	OIL
56	6451.5	2130	31.0	12.9	55.9	OIL
57	6452.5	645	32.9	10.8	55.0	OIL
58	6453.5	384	36.9	8.0	63.8	OIL
59	6454.5	670	36.2	10.2	61.8	OIL
60	6455.5	2820	34.9	10.3	58.7	OIL
61	6456.5	5300	35.3	11.0	60.8	OIL
62	6457.5	1180	36.1	0.0	69.0	WATER
63	6458.5	4970	36.9	0.0	71.2	WATER
64	6461.5	4570	36.1	0.0	61.8	WATER
65	6462.5	4540	36.8	0.0	68.1	WATER
66	6463.5	822	34.8	0.0	69.8	WATER

COMPLETION COREGRAPH

PERMEABILITY — 100000 MD 10   POROSITY — 40.0 % 13.0   DEPTH 1:30 F   LITHOLOGY

CORE SATURATIONS (% Pore Volume)
Oil Saturation 0 % 100
Water Saturation 100 % 0

FIGURE 6.10
A typical core analysis log. (Courtesy of Core Laboratories, Inc.)

where

r_e = the inner radius

$_{RW}$ = the outer radius

P_e = the pressure at r_e

P_w = the pressure at r_w

TABLE 6.3

Klinkenberg Corrections for Gas Flow

Air Permeability, K_A millidarcies	Klinkenberg Permeability, K_1 millidarcies	Ratio K_1/K_A
0.18	0.12	0.66
1.00	0.68	0.68
10.00	7.80	0.78
100.00	88.00	0.88
1000.00	950.00	0.95

Courtesy Core Laboratories.

6.8   Problems Associated with Core-Derived Data

When core-derived measurements are correlated with wire line measurements, a length of core equal to the diameter of the logged volume must be averaged, preferably by a weighting method which will match the sensitivity variation of the wireline measurement.

Core-derived measurements are subject to many of the same type problems that other measurements are, and some peculiar to coring. They are affected by the small sample size, which results in a high statistical variation. The core is subject to damage during acquisition, retrieval, storage, transportation, and processing. Unconsolidated cores are especially difficult to handle. There is a difficulty reproducing downhole environments for laboratory use. Saturation determinations are sometimes difficult to make accurately, because the sample has been flushed with drilling mud during drilling and retrieval, gases will have expanded and liquids will have contracted during the trip to the surface because of severe temperature and pressure changes. Moreover, coring methods are slow and expensive. They cost approximately ten to one hundred times wire line information, per unit of information.

In spite of the problems, porosity determinations are usually good. This is partly because we have a history of coring experience for more than a thousand years. Also, a good core laboratory will be aware of and take steps to correct these problems.

Figure 6.10 shows a typical petroleum core analysis report.

6.9   Cuttings Samples

Cuttings samples are the material cut away by the drill bit and carried to the surface by the drilling mud, where they are collected. They are a valuable

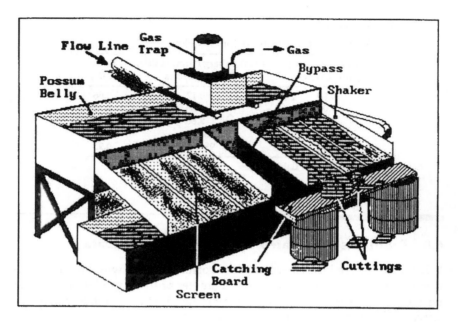

FIGURE 6.11
One of many kinds of shale shakers for collecting cuttings during drilling.

source of information if their shortcomings are recognized. They may be used to recognize changes of rock type and redox state. During drilling for petroleum, they are used to detect the presence of hydrocarbon. In mineral work, they are a good source for stratigraphic determinations. Figure 6.11 shows a shale shaker, one of many types used to collect cuttings samples during drilling. This model eliminates most of the shale which is mixed with the cuttings.

Cuttings samples suffer from the small sample size, high contamination with drilling mud, severe damage in the drilling process, some of their components, especially the fines, being washed away, and uncertain depth of their source.

6.10 Sidewall Coring

Sidewall cores can be taken with wire line equipment. The samples are very small and they are often badly damaged. They are, however, good for remedial and confirming purposes. See Figure 6.4 for a drawing of a sidewall sample gun.

Sampling can also be done with other, less often used methods. A drop corer may be used in soft zones. This device is an open-ended tube which is dropped with the open end down. The inertia of the device drives it into the material to be sampled. It is the retrieved with a wire line.

6.11 Fluid Sampling

Formation fluid samples may be taken in several ways. The driller can often obtain a sample of the formation water by placing packers in the well above and, sometimes below the permeable zone and flowing or pumping the fluid from the formation. The first fluid is from the invaded zone. Later flow is from the undisturbed zone. This method is common in water well drilling.

A second method is the production-type drill stem tests. These are done in a manner similar to the water sampling described in the previous paragraph. The goal here, however, is to sample the hydrocarbon content. Careful records are kept of the fluid amounts and types, as well as continuous records of the pressures.

A third method is the wireline fluid tester or formation tester. This is a tool sent down on the logging wire line. The fluid sample consists mostly of fluid from the invaded zone. Some contractors claim otherwise, however. It operates in moderate to high porosity zones. Pressure records are kept during the fluid flow. Figure 6.5 shows a drawing of a fluid sample taker. The wireline formation tester is accurately positioned with information from the open hole logs, especially the SP and the gamma ray logs. When the tool is in position, a pad is moved out toward the wall of the hole, usually using the hydrostatic pressure of the mud column for actuating energy. When the pad is sealed against the formation, a valve is opened by a signal from the surface to allow the formation fluid to flow into the tool reservoir (or one of the reservoirs, if the tool has more than one). Sample sizes range from about 10 cc to more than 2 liters. A surface recording is made of the sampling pressure during the fluid flow. When the reservoir is full, the valve is closed and the shut-in pressure is recorded. The pad is then retracted and the hydrostatic pressure of the mud column is recorded.

This wireline fluid tester service supplies data concerning production, flow rates, water cut (rather doubtful value, since the fluid is from the invaded zone), gas:oil ratio (again, of questionable value), and permeability.

7

Introduction to Electrical Resistance and Resistivity

7.1 Resistance and Resistivity

Resistivity methods constitute a major group of measurements in petroleum formation evaluation. They determine an important physical property of a material. For many years, resistivity logs were the only logging method which would directly and quantitatively respond to the presence of hydrocarbon in the formation material. Formation electrical resistivity, R, is increased due to the presence of hydrocarbon within the zone. In addition, resistivity measurements tell us something about the clay/shale content, formation waters, structure, stratigraphy, and age, among other things. Many surface electrical measurements and a large percentage of downhole geophysical measurements are some form of quantitative resistivity measurement. Engineering and scientific measurements demand quantification.

Electrical resistance, r, measurements are also made downhole. These measurements quantitatively tell little about the formation zones. They are, however, valuable for stratigraphic determinations in small diameter bore holes in low salinity muds.

Both types of systems are useful and in common use. Both parameters, R and r, can be very useful, but they must not be confused. Resistance is commonly used in sedimentary mineral exploration because the fluid content of a sand is not very important. The single point resistance curve, however, is an excellent stratigraphic curve. It is sensitive to the sands, the degree of shaliness, and the redox state. *It must not be called "resistivity," however.*

7.2 Definitions

Electrical resistance, r, is the impedance to the flow of electrical charges (electrons or ions). Its reciprocal is conductance, c. It is a property of a unique geometry of a unique sample of material. The measured resistance, r, is peculiar to that sample and that sample only. It depends upon the shape of the sample, as well as the type material, temperature, and pressure.

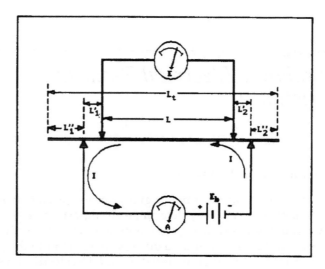

FIGURE 7.1
A simple circuit to flow electrical current through a wire.

Electrical resistivity, R, is a fundamental property of a material. Its reciprocal is conductivity, C. The property, R (or C) is independent of the size and shape (the geometry, G) of the sample of material. Resistivity, R, and conductivity, C, depend only upon the type of material, its temperature and (usually to a lesser extent) its pressure.

7.2.1   Resistance

Resistance measurements require that an electrical current flow through the material to be measured. To do this we must put an electrical field across the sample to be measured, to cause the measure current to flow through it, as shown in Figure 7.1. This shows a battery, E_b, connected to a length, L_t, of wire. The battery is an energy source and will set up a voltage field along the wire from the negative end toward the positive end. This voltage field will cause a flow of the electrons in the wire from the negative end to the positive. This is a flow of electrical current. An ammeter, A, will indicate the flow of current, I, through the system loop. The current will be the same at any place in this simple loop. Thus, the ammeter will measure the amount of current through this wire. Note that the same current flows through L, L_1', and L_2', but no current flows through L_1'' and L_2''.

If a voltmeter is placed across the wire at points P_1 and P_2, we can measure the voltage drop, E, between the two electrodes of the voltmeter (distance L). Note that the distance L does not include L' and/or L''. The current, I, however, is the same anywhere within the current loop, even in L', outside of L.

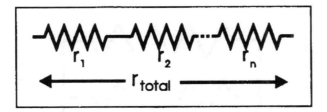

FIGURE 7.2
Series resistances.

For any system, the ratio of the voltage to the current is a constant. It is the resistance, r, in ohms, of the sample. This is Ohm's Law:

$$r = \frac{E}{I} \qquad (7.1)$$

We can calculate the resistance of this segment of wire, L. If we were to include L_1' and/or L_2', the voltage drop, E, would be greater, but the current, I, would be the same. Thus, the resistance of the longer length, $L + L_1'$ and $L + L' + L_2'$ would be greater.

7.2.2   Resistivity

The common symbol for a resistance is the resistor (Figure 7.2). We learned in elementary physics that resistors can be added in series, resulting in

$$r_{total} = r_1 + r_2 + L + r_n \qquad (7.2)$$

Also, when the resistors are combined in parallel (Figure 7.3) that:

$$\frac{1}{r_{total}} = \frac{1}{r_1} + \frac{1}{r_2} + L + \frac{1}{r_n} \qquad (7.3)$$

These elementary principles also apply if a block of material is substituted for the resistor. As a matter of fact, that is exactly what a commercial resistor often is, a block or rod of carbon. If the material of the block is held constant (Figure 7.4), but the length, L, and/or the cross sectional area, A, is changed, the total resistance of the block is equal to the length, L, and the reciprocal of the cross section, $1/A$, and to some fundamental, constant property, R, a proportionality constant:

$$r_{total} = R \frac{L}{A} \qquad (7.4)$$

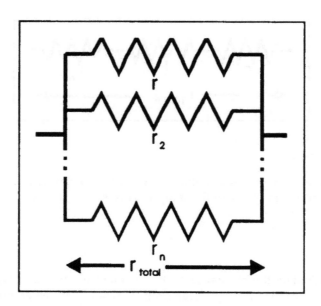

FIGURE 7.3
Parallel resistances.

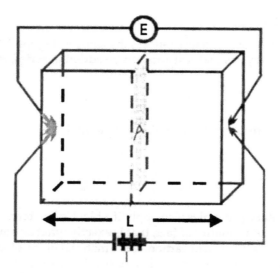

FIGURE 7.4
Determining the electrical resistance of a block of material.

The proportionality factor, R, is the resistivity. It is a property of the material, whereas the resistance, r, is a function of the shape of the material and its resistivity. If we solve for R,

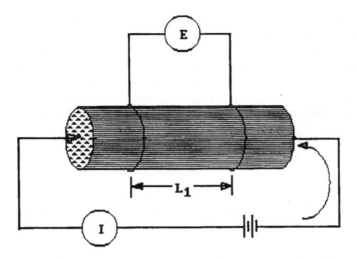

FIGURE 7.5
The determination of the resistivity of a core sample.

$$R = r\frac{A}{L} \qquad (7.5)$$

This is precisely how the resistivity of a core is determined, as shown in Figure 7.5. The circular cross section of the core is A and the distance between the measure electrodes is the value of L. The equation may be generalized:

$$R = G_r r \qquad (7.6)$$

where G_r is a constant which describes the geometry of a particular system; the geometric constant.

This explanation is important because the terms *resistance*, r, and *resistivity*, R, are often confused. Many times, when the single point resistivity system is used, the logged curve is labeled "resistivity" when it is really *resistance* which is presented.

Electrode-type resistivity and resistance equipment (i.e., single point, normals, focussed electrode systems, micro-resistivity) *all* measure the voltage of the measure electrode. The circuit constants (i.e., current, I) determine the resistance:

$$r = \frac{E}{I} \qquad (7.1)$$

This, in turn, *can be* converted to resistivity, R, if the geometry of the system is known and applied:

$$R = Gr \qquad (7.7)$$

In the earlier example, in this chapter, the geometrical constant used the dimensions of the sample, A/L. This relationship identifies a fundamental property of the material which can be used for many purposes (material identity, fluid content, and many others).

Resistivity is very important in petroleum work because that field is interested in the resistivity contrast between the formation fluids. And, because of the size of the petroleum targets, high resolution is of secondary importance.

7.2.3   Units

In the petroleum business, the units of G, will always be square meters per meter. Thus, the units of R are ohms meters squared per meter which is commonly shortened to ohmmeters. Surface geophysicists frequently use ohm-centimeters and ohm-feet. R is a physical property of the material, whether it be water, core rock, wire, or any other material. It is independent of the geometry of the sample. You will find many tables of resistivities of materials in the handbooks. See also Table 3.3 in Chapter 3.

The reciprocal of r is c, the conductance. Its unit is the Sieman or the mho (an older term, but still used). The reciprocal of resistivity, R, is conductivity, C. Its unit is Siemens per meter or mhos per meter. Since this is a large unit, the common ones are milli-Siemens per meter and millimhos per meter. Most (but not all) electrical measurements are scaled in ohmmeters (Ωm). All electromagnetic systems (such as the induction log) measure conductivity and may be scaled and/or calculated in conductivity units (most of the time the conductivity units are electronically reciprocated and displayed in ohmmeters).

7.3   Formation Resistivity

Now, let us suppose that we have a cubic block of quartz whose length is 1 meter (1 m) and whose cross-sectional area is one square meter (1 m^2), as shown in Figure 7.6. We might call this a block of quartz sandstone whose porosity is zero. Its volume, V_q, is 1 m^3 (after Doveton, 1986).

If we measure the resistivity of this block, we find that the current is very small and, therefore, the resistance and resistivity are very high. The resistivity, R_q, is between 10^{10} and 10^{15} ohmmeters. In fact, it is so high that our logging systems, which are built for lower resistivity ranges, will read R_q as apparently infinite. The resulting currents are on the order of 10^{-10} to 10^{-15} amperes (per volt), or virtually zero for our purposes.

We can picture this block as a stack of quartz rods, each with an area of 1 square centimeter (1 cm × 1 cm, 1 cm^2) and 1 m long. The resistance, r_q,

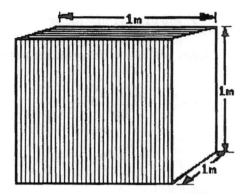

FIGURE 7.6
A cube of quartz with dimensions of one meter per side.

of each rod is in parallel with that of each other rod. There will be 10,000 (100 × 100) of them. The resistance, in ohms, of each quartz rod, r_q, is

$$r_q = R_q \frac{L}{A} = R_q \frac{1}{0.0001} = 10^{+4} R_q \tag{7.8}$$

Therefore, the conductance of the quartz block, $1/r_o$ (c_o), is

$$\frac{1}{r_o} = \frac{1}{r_{q1}} + \frac{1}{r_{q2}} + L + \frac{1}{r_{q10,000}} = \frac{10^4}{r_q} = \frac{1}{R_q} \tag{7.9a}$$

or, in terms of conductance

$$c_o = c_{q1} + c_{q2} + L + c_{q10,000} = 10^4 c_q \tag{7.9b}$$

That is, for the 1 m × 1 m × 1 m block, numerically:

$$r_q = R_q \tag{7.10}$$

If we examine a block of salt water of the same dimensions, it will have a low resistivity, R_w (between 10^{-2} to 50 Ωm). This block has a porosity of 100%. Its resistance, r_w, is, numerically:

$$r_w = R_w \frac{L}{A} = R_w \frac{1}{1} = R_w \tag{7.11}$$

If we drill holes of 1 cm² area through the quartz block (or remove rods) and fill each hole with salt water, the volume, V_g, of each hole will be:

$$V_s = (0.01\,m)^2 (1\,m) = 10^{-4}\,m^3 \qquad (7.12)$$

The total volume of n holes will be nV. The resulting porosity, ϕ, will be:

$$\phi = \frac{nV}{V_q} = \frac{n\left(10^{-4}\,m^3\right)}{1m^3} = 10^{-4}\,n \qquad (7.13a)$$

$$n = 10^4\,\phi \qquad (7.13b)$$

The resistance, r_w of each hole will be:

$$r_w = R_w \frac{L}{A} = R_w \frac{1}{10^{-4}} = 10^4\,R_w \qquad (7.14)$$

If there is only 1 hole in the block, the resistance of the block, r_o, (in ohms) will now be

$$\frac{1}{r_o} = \frac{10^4 - 1}{r_q} + \frac{1}{10^4\,R_w} \qquad (7.15)$$

Since, $r_q \simeq 10^{15}$ ohms, for our purposes,

$$\frac{10^4 - 1}{r_q} \approx 0$$

then

$$r_o = 10^4\,R_w \; ohms. \qquad (7.16)$$

With n holes,

$$\frac{1}{r_o} = \frac{10^4 - n}{r_q} + \frac{n}{10^4\,R_w} \qquad (7.17)$$

and

$$r_o = \frac{10^4\,R_w}{n} \qquad (7.17a)$$

Then, from Equation 7.5a

$$R_o = r_o \frac{A}{L} = \frac{10^4 R_w}{n} = \frac{R_w}{\phi}$$ (7.18a)

and, combining with Equation 7.12b

$$\frac{R_\varrho}{R_w} = \phi^{-1}$$ (7.18b)

The exponent, −1, describes the geometry of this simple pore space. We can substitute a general geometric value, m, in its place:

$$\frac{R_\varrho}{R_w} = \phi^{-m}$$ (7.19)

This is Archie's porosity equation. Note that virtually all of the measure current flows through the water-filled pore spaces. This current is on the order of amperes per volt. That flowing through the quartz is on the order of 10^{-15} amperes per volt. Thus, the resistivity measurement tells us something about the pore space (geometry and fluid) but essentially nothing about the character of the rock material, itself.

In actual practice, the pore paths are many times more tortuous than the straight hole. That is, they bend, branch, expand, and constrict. Therefore, the actual value of the resistance is determined by the real path length and cross section and not by the length of the bulk sample ($m > 1$).

Regardless of the resistivity of the fluid:

$$F_r = \frac{R_\varrho}{R_w}$$ (4.11)

F_r is the "Formation Resistivity Factor" (or simply the formation factor) and is a property of that sample or type of sample.

7.4 Cementation Exponent

In 1942, Gerald Archie, of Shell Oil Company, found empirically that the formation factor is related to the porosity:

$$F_r = a \phi^{-m}$$ (4.31a)

where "m" is called the cementation exponent. It is an exponent describing the pore geometry. We have seen earlier, in Equation 4.9, that "m" is related to the tortuosity of the pore space.

$$t = \frac{L_e}{L_b} \tag{4.9}$$

Pirson (Pirson, 1935) described the tortuosity, t, as:

$$t = \left(\frac{L_e}{L_b}\right)^2 \tag{7.20}$$

where L_e is the effective length of the pore channel in a sample and L_b is the straight line length of the same pore. Lynch (Lynch 1962) defined tortuosity as:

$$R_o = \left(\frac{L_e}{L_b}\right)^2 \frac{1}{\phi} R_w \tag{7.21}$$

Since

$$\frac{R_o}{R_w} = \phi^{-m} \tag{7.22}$$

Then

$$m = 1 + \frac{2\log\left(\frac{L_e}{L_b}\right)}{\log\phi} \tag{7.23}$$

Archie found that the value of "m" for an unconsolidated sand is about 1.3. The value of "m" is regulated almost entirely by the degree of tortuosity of the pore space. This is the complement of the rock matrix geometry, which is controlled by the rock texture. The values shown in Table 7.1 are usually used.

If the value of "m" is plotted against the value of "F," using Archie's porosity equation (in which the factor "a" is always equal to 1), Equation 4.31a, the plot is shown in Figure 7.7.

Winsauer (Winsauer, 1952) suggested that Archie's equation, in actual practice, needs a factor "a" which is frequently different from 1. This helped the problems of getting a better relationship between porosity and the formation factor. The modified Archie relationship of Equation 4.31a is:

TABLE 7.1

Approximate Values of the Exponent "*m*"

m	Porous Rock
1.3	Unconsolidated
1.4–1.5	Very slightly cemented
1.5–1.7	Slightly cemented
1.7–1.9	Moderately cemented
1.9–2.2	Highly cemented
2.0–2.7	Carbonates

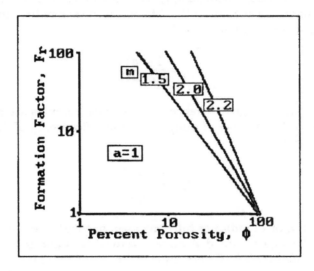

FIGURE 7.7
A plot of formation factor vs. porosity.

$$F_r = a\phi^{-m} \qquad (4.31a)$$

If $a \neq 1$, this relationship fails at high and low values of porosity, however. This indicates that "*a*" (if, indeed, it is a valid constant) is probably some variable function of porosity, rather than a constant. In most sands (in the ranges where we normally work), the relationship works well when $a = 1$. One common exception is the Humble relationship for Texas Gulf Coast sands:

$$F_r = 0.62\phi^{-2.15} \qquad (7.24)$$

This is an empirically determined relationship and should *not* be used without *good evidence* that it applies.

We must remember that Archie's and Windsauer's relationships are empirical and were derived to fit their numbers. This does not detract from the value and usefulness of the equations. Efforts are being made to explain the ϕ:R relationships. They invariably find a strong influence of pore geometry or "tortuosity" on the values of "m". This is because a major change of porosity is usually accompanied by a change of the geometry of the pore space (tortuosity). "m" (or some function of "m") is not a constant over a wide range of porosities. Various characteristics of the formation materials have been correlated with "m" (Doveton, 1986), using the Spearman rank correlation coefficient, r_s. This is a function of the number of samples and of the average of the differences in rank of the samples. The value of r_s may vary from +1 to –1:

$$r_s = \frac{1 - 6d^2}{n(n^2 - 1)} \tag{7.25}$$

where d is the average difference in rank when samples are ordered in rank, and n is the number of samples. The significance, T_s, of the value of r_s can be tested:

$$T_s = \frac{r_s(n-2)}{(1 - r_s^2)^{1/2}} \tag{7.26}$$

A value of r_s of +1 indicates full agreement or trend. And, –1 indicates a diametrically opposite trend. A value of zero would indicate no trend. Tortuosity was measured by determining the transit times of ions through the pore spaces of core samples. Winsauer found a Spearman rank coefficient of +0.72 (using his measures of tortuosity and "m"). When "m" was correlated with absolute permeability, k_a, the Spearman coefficient was found to be –0.64. This is a significant correlation. It is consistent with the intuitive notion of geometrical control of fluid flow.

The correlation with the standard deviation of grain size was not as good, but still significant. It was found to be +0.34. The correlation with grain roundness was found to be –0.34. The correlation of "m" with skewness (–0.22), mean diameter (+0.20), and packing index (+0.19) were low enough that they probably are not significant.

Raiga-Clemanceau suggests a porosity relationship which takes permeability (a function of tortuosity) into account:

$$F_r = \phi^{-m_k} \tag{7.27}$$

where (k is the permeability in millidarcies)

$$m_k = 1.28 + \frac{2}{\log_{10} k + 2} \tag{7.28}$$

7.5 Rock Texture — Sandstones

It has been demonstrated (Pettijohn et al., 1972) that the textural properties of rock which have been discussed, are not independent but are controlled by cycles of weathering and the physics of transport hydraulics and deposition. This can be thought of in terms of "textural maturity" (Folk, 1951). The samples in one test gave average values of "m" of 1.87 for immature samples, 1.78 for sub-mature samples, and 1.76 for mature samples. They found a lack of coherent patterns which was probably because these trends were obscured by the compaction and diagenesis mechanisms.

An analysis of petrophysical and petrographical data from cores of the Skinner sandstone (Pennsylvanian) from South-east Kansas has been done (Doveton, 1986). Indications were that this was an alluvial channel fill. Correlations with "m" were weak for grain size, sorting, cement proportion, and mica content. A comparison of the core description of bedding structure, however, showed a distinct pattern. Ripple bedded zones had a value of "m" of 2.20, while the cross bedded zones had 2.05. The permeability was 25 md for the ripple bedded zones and 40 md for the cross bedded zones.

The value of "m" can be calculated using Archie's equation and logged porosity values and plotted as a log (Doveton, 1986). This is a technique for determining textural or facies changes in sandstone type reservoirs. In practice, the value of "m" is highly variable and determined by the fracture size, orientation, and the nature of their contained fluids. Because of the way it measures, the induction log will have a tendency to ignore vertical fractures but will detect horizontal ones when they are filled with water. By contrast, the Laterolog resistivities will tend to be lower than the induction log values in vertical fracture systems. The values of "m" in the Arbuckle limestone (Cambro-Ordovician) in Kansas average very close to 2.0, with minor excursions due to vugs and fractures. The Pennsylvanian, on the other hand, shows wide variations, due to dissolution of ooids and high tortuosities.

7.6 Rock Texture — Carbonates

In carbonates, Archie noted that his relationship held also, but with a greater scatter of values. The meaning of "m" in carbonates is more difficult to determine than in sandstones. Archie's porosity relationship best typifies carbonates with intergranular and inter-crystalline types of porosities.

Chombart (Chombart, 1960) reported values of "m" generally between 1.8 and 2.0 for crystalline and granular carbonates, 1.7 to 1.9 for chalky limestones, and 2.1 to 2.6 for vugular carbonates. Suau and Gartner (1980) report that the value of "m" will fall to values near 1.4 in fractured systems.

In fractured carbonate systems, a parallel model has been proposed, where cells of limestone with matrix porosity, ϕ_{ma}, are in parallel with planar fractures, ϕ_f. If the model is saturated with water, the total porosity, ϕ_t, becomes

$$\phi_t = \left[\left(1 - \phi_f \right) \phi_{ma}^m + \phi_f \right]^{1/M} \tag{7.29}$$

where

ϕ_f = the porosity of the fracture system

ϕ_{ma} = the porosity of the rock matrix (except the fractures)

ϕ_t = the total porosity

m = the value of the cementation exponent of the matrix

M = the value of the of the cementation exponent of the total rock

7.7 Salinity

The salinity of a water/electrolyte mixture is important because the flow of ions of the salts and metals in the formation water is the medium of electrical current flow in the formation materials. If a metal electrode is put into water (or any other ionizing medium), ions of the metal will immediately start to enter the water. The rate of entry will drop off as the water surrounding the electrode becomes saturated with metal ions. This is illustrated in Figure 7.8.

If, however, two electrodes are immersed in the water and are connected together through a voltage source, such as a battery, as shown in Figure 7.9, a voltage field will be set up in the water. The ions will begin to flow, the positive ions (cations) will be attracted to the negative electrode (cathode). The negative ions or anions will begin to flow toward the positive electrode (anode). The flow of ions is an electrical current flow. When the cations reach the cathode, they will take an electron from it and be neutralized. Similarly, when the anions reach the anode, they will give up an electron to the anode and be neutralized. These are redox reactions (see Figure 7.9).

The salinity of a solution may be estimated from the value of the electrical resistivity of the solution. This calculation *must be made using the resistivity value of R_w at 75°F (24°C)*.

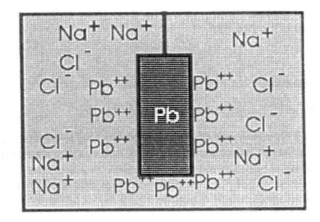

FIGURE 7.8
Population of ions around a lead electrode in water.

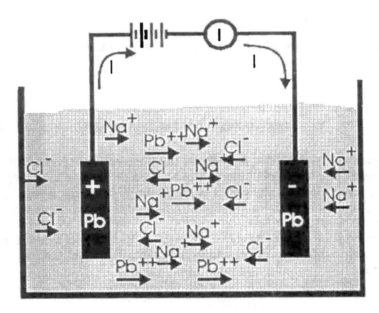

FIGURE 7.9
The migration of ions in an electric field.

$$ppm_{eNaCl} = 10^x \qquad (7.30)$$

where

$$x = \frac{3.562 - \log(R_w - 0.0123)}{0.955} \qquad (7.30a)$$

Metallic solids exhibit electronic conduction. The results are the same, as far as our logs are concerned.

7.8 R_T VS ϕ Crossplot (The Pickett Plot)

Richard Pickett suggested using a plot of the Archie equation for porosity to determine the probable value of some of the unknowns and for making corrections when working with resistivity logs. The method is a valuable illustration of the resistivity relationships. Archie's porosity equation states:

$$R_o = R_w \phi^{-m} \tag{7.31}$$

$$R_o = R_t S_w^n \tag{7.32}$$

Combine these, solve for R_t, and the take the logarithm:

$$R_t = R_w S_w^n \phi^{-m} \tag{7.33}$$

$$Log(R_t) = \log(R_w) + n\log(S_w) - m\log(\phi) \tag{7.34}$$

Equation 7.34 is a linear equation of the slope intercept form. Plotting it will result in a family of curves which are straight lines on a log/log grid. The y-axis is $\log(R_t)$, the x-axis is $\log(\phi)$, the y-intercept is $\log(R_w) + n \log(S_w)$ and the slope is $-m$. See Figure 7.10. Iso-saturation lines are parallel to the 100% saturation line.

The value of m is the negative slope value and may be determined by using values from one of the parallel saturation lines (i.e., the $Sw = 100\%$, R_o line):

$$-m = \frac{\Delta y}{\Delta x} = \frac{\log R_{t1} - \log R_{t2}}{\log \phi_1 - \log \phi_2} \tag{7.34a}$$

If S_w is set equal to 1 (100%), then

$$R_t = R_o \tag{7.35}$$

and the S_w term becomes equal to zero. Under these conditions, the trend will define values of R_o. The equation is now Archie's porosity equation. If S_w and ϕ are both equal to 1, then both of their terms become zero and

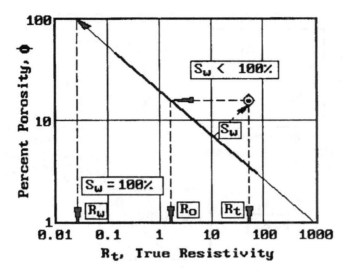

FIGURE 7.10
A crossplot of R_t vs. ϕ.

$$R_t = R_o = R_w \qquad (7.36)$$

In this way, the value of R_w for the system may be found.

If the value of S_w is less than 1 at any value of ϕ, the R_t value will be displaced at a constant porosity, ϕ, along the y or R_t direction, by a value of

$$S_w = \left(\frac{R_o}{R_t}\right)^{\frac{1}{n}} \qquad (7.37)$$

Since both R_t and R_o are now known, the value of S_w may be calculated for any value of n.

7.9   The Nonlinear (Hingle) Crossplot

A similar plot, known as the nonlinear or Hingle plot was proposed by Thomas Hingle. This diagram also plots the value of R_t against porosity, ϕ, or a function of porosity. The value of this plot is that one can use it to begin to identify the rock matrix type, in addition to estimating a good value of water saturation, S_w.

If the formation factor, F_r, is plotted against porosity, ϕ, the trend is nonlinear. See Figure 7.11a. This response can be made linear by using values of:

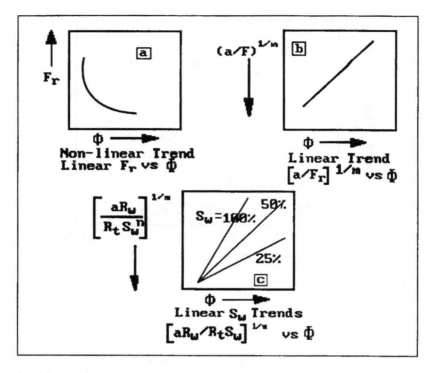

FIGURE 7.11
The process of obtaining the non-linear plot of resistivity vs. porosity.

$$\left(\frac{a}{F_r}\right)^{\frac{1}{m}} \tag{7.38}$$

for the y-axis, in place of F_r. This is shown in Figure 7.11b. The saturation values can also be incorporated by using a value which is also a function of F_r and m, as well as saturation, S_w:

$$\left(\frac{aR_w}{R_t S_w^n}\right)^{\frac{1}{m}} \tag{7.39}$$

This is shown in Figure 7.11c. The y-axis is scaled in values of R_t in a negative direction (increasing downward). The grid is a function of "m". Trends of S_w values will be straight lines. All of the S_w trend lines will converge at $R_t = \infty$. At a value of $R_t = \infty$, the porosity is zero and the function of porosity will define the character of the solid-rock matrix. A commercial sample of this is shown in Figure 7.12.

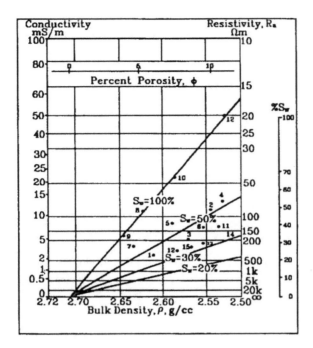

FIGURE 7.12
An example of a plot of formation resistivity against bulk density and porosity on a non-linear or Hingle plot.

8

Introduction to Radioactivity

8.1 Definitions and Terms

Radioactivity is the emission of radiant energy. It is also the property of certain elements of spontaneously emitting radiant energy, by the disintegration or change of the nuclei of their atoms. This radiant energy may be either electromagnetic wave radiation or it may be particles.

The inverse square law states that the radiation intensity per unit area varies inversely with the square of the distance from the source of radiation. It applies strictly to point source radiation:

$$\frac{d_1^2}{d_2^2} = \frac{I_2}{I_1} \qquad (8.1a)$$

or

$$I_2 = I_1 \frac{d_1^2}{d_2^2} \qquad (8.1b)$$

where

    d = distance
    I = intensity

The disintegration relation describes the disintegration of radioactive atoms, or their activity:

$$\frac{dN}{dt} = -\Lambda N \qquad (8.2a)$$

$$\int_{N_0}^{N} \frac{1}{N} dN = -\Lambda \int_0^t dt \qquad (8.2b)$$

$$\therefore N = N_0 e^{-\Lambda t} \qquad (8.2c)$$

where

N = the number of events

N_0 = the original number of atoms

m = the disintegration or activity constant (the number of disin-
    tegrations per second) (also called the probability constant)

t = time in seconds

An example of this is shown in the uranium decay series, published by the
U.S. Department of Energy and shown as Figure 8.1.

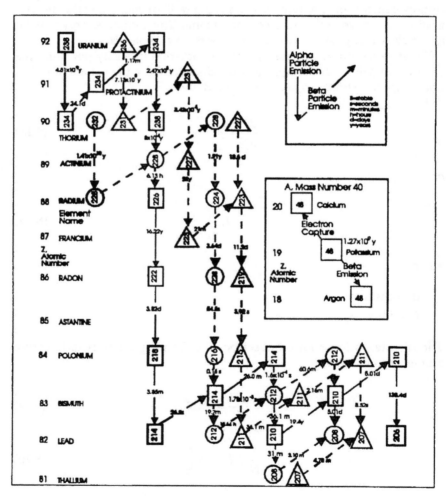

FIGURE 8.1
The decay progression for uranium, thorium, and their daughters.

Halflife, $t_{1/2}$, is the length of time required for 50% of of an exponential function to occur. In this case it is the length of time for 50% of radioactive isotope nuclei to disintegrate:

$$t_{1/2} = \frac{\ln(2)}{\Lambda} \tag{8.3}$$

Einstein's equations for photoelectric interaction between a photon and an orbital electron,

$$h\nu = \frac{1}{2}mv^2 + E\omega \tag{8.4a}$$

mass to energy conversion,

$$E = mc^2 \tag{8.4b}$$

and kinetic energy,

$$E = m_o c^2 \frac{1}{\left(1 - \frac{v^2}{c^2}\right)^{\frac{1}{2}} - 1} \tag{8.4c}$$

describe most of the radioactive reactions, where

c	= the velocity of light in a vacuum
h	= Planck's Constant (6.6×10^{-27} erg seconds)
ν	= the frequency of the photon
m	= the mass of the escaping electron
m_o	= the rest mass of the electron
v	= the electron velocity
$1/2\, mv^2$	= the kinetic energy
$E\omega$	= the energy necessary to release the electron (the work function)

An electron volt (eV), is defined as the energy required to cause one electron to fall through a field of one volt.

$$1 eV = 1.60219 \times 10^{-12} erg = 3.8 \times 10^{-20} cal = 1.78253 \times 10^{-33} g \tag{8.5a}$$

$$1ev = \frac{1}{8.058 \times 10^3 \, \text{Å}} \tag{8.5b}$$

where

Å = Angstrom units, a measure of wave length.

The unit, Å, is used most often in X-ray technology and mineralogy:

$$1\text{Å} = 10^{-8} cm \tag{8.6}$$

Charge is in electrostatic units (esu). An electrostatic unit is defined as a charge resulting in a force of one dyne repulsion from an equal charge 1 centimeter away.

A Curie (Ci) is a unit of radioactive strength or activity:

$$1Ci = 3.7 \times 10^{10} \, d/s = 3.7 \times 10^{10} \, Bq \tag{8.7}$$

where

d/s = disintegrations per second

Bq = Bequerels

The scattering of photons and radioactive particles is the result of inter-action between the atoms of the material being transversed and the incident particle, resulting in a change of velocity of and a transfer of energy between the two interacting entities. The intensity, I, of a beam of radiation, after scattering through a medium, is a function of the original intensity, I_o, the density of the scattering medium, ρ, and the mass attenuation coefficient, μ (Figures 8-2 and 8-3):

$$I = I_o e^{-x\rho\mu} \tag{8.8}$$

where

x = distance through the target material

ρ = the bulk density of the target

μ = the mass absorption coefficient

A Roengten unit (R) is a radioactivity health unit of ionization, in air:

$$1R = 6.76 \times 10^{10} \, eV \tag{8.9}$$

An atomic mass unit, AMU, is the mass of a hydrogen atom:

$$1AMU = 931 \, MeV \tag{8.10}$$

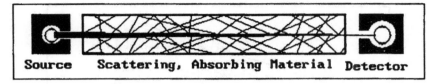

FIGURE 8.2
Attenuation by scattering.

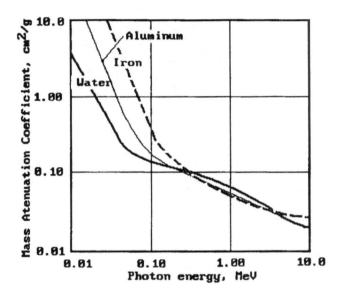

FIGURE 8.3
The mass attenuation coefficient, μ, vs. atomic number, Z.

8.2 Particle Descriptions

8.2.1 Electrons, Beta Particles, and Positrons

Beta particles are negative electrons. When they have been ejected from an atomic orbit by interaction with a photon, they are called photoelectrons. In a stable atom, they orbit the nucleus and are responsible for the chemical characteristics of the atom. Photoelectrons or beta particles can interact with the orbital electrons, as shown in Figure 8.4. Beta particles can interact with the nucleus, also, in a Bremsstalung reaction, shown in Figure 8.5. Free electron currents are the usual electrical conductive medium in metallic

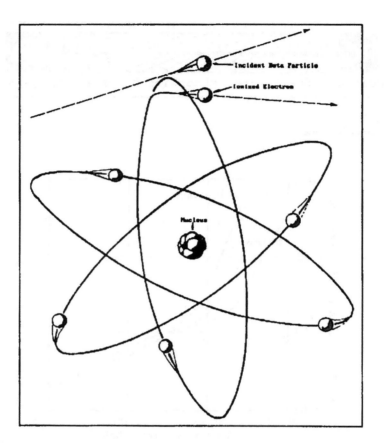

FIGURE 8.4
Beta particle interaction.

materials. The electron rest mass is approximately 1/1800 AMU. It has a
charge of –1 esu. Because of its charge and small mass, it is not very penetrating. It can be blocked by a sheet of aluminum foil. The wave length of
an electron is given by:

$$\lambda = \frac{h}{mv} \tag{8.11}$$

where h is Planck's Constant, m is the mass of the electron, and v is its
velocity.

A positron is a positively charged electron or an antielectron. It has a
mass of 1/1800 AMU and a charge of +1 esu. It has a short half life because
it will quickly react with a negative electron with the annihilation of both
particles and the emission of energy at 511 kev.

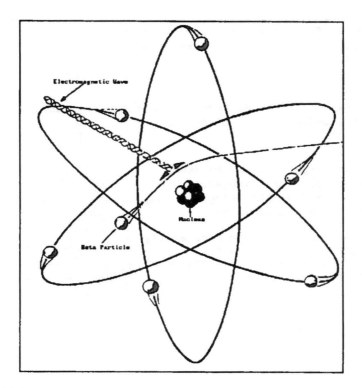

FIGURE 8.5
Bremstrahlung production.

8.2.2 Neutrons

A neutron is a neutral particle with a mass of approximately one atomic mass unit (AMU). It is very penetrating because of its zero charge. Refer to Figure 8.6 and Table 8.1. Its energy is manifested in its mass and velocity. It can be thought of as a combination of a (positive) proton and a negative electron. This text will treat it as a solid, neutral, elastic particle. Free neutrons are seldom found naturally. Neutrons originate from the breakup of some nuclei by artificial means. A few neutrons originate from uranium, but they are very few. In some deposits, there is evidence that enough ^{235}U existed, at one time in the distant past, to emit neutrons and sustain fission. Apparently, all such natural deposits underwent such spontaneous fission many millions of years ago.

A free neutron has a half life of 10.6 minutes and decays to a proton with the emission of a beta particle:

$$n \rightarrow 1H + {}^{-1}\beta \tag{8.12}$$

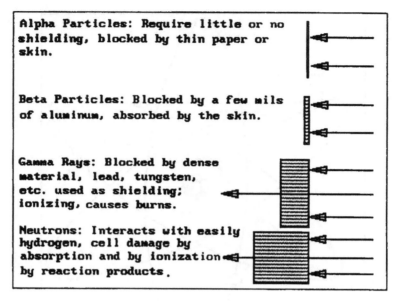

FIGURE 8.6
Penetrating abilities of various radioactive particles.

TABLE 8.1

Elementary Atomic Particles

Particle name	Symbol	Rest mass (AMU)	Charge (esu)	Position	Description
Photon		0	0	Free	Electromagnetic unit particle
Neutron	n	1	0	Nuclear	Neutral particle
Electron	e^-,β^-	0.0006	−1	Free & orbital	Beta particle
Positron	e^+,β^+	0.0006	+1	Nuclear	Positive electron
Proton	$p,{}_1^1H^+$	1	+1	Nuclear	Hydrogen nucleus
Deuteron	${}_1^2H^+$	2	+1	Nuclear	Deuterium nucleus
Triton	${}_1^3H^+$	3	+1	Nuclear	Tritium nucleus
Alpha	${}_2^4He^{++a}$	4	+2	Nuclear	Helium nucleus

The proton, $^1H^{+1}$, will quickly capture any free electron and become a stable hydrogen atom.

Neutrons interact with atomic nuclei by direct collision. Interaction can be elastic, with the neutron imparting some of its energy to the atom, which recoils, if it has a low atomic number. Recoil is negligible if the atom has a high atomic number. The neutron will retain the remainder of the energy. See Figure 8.7.

The collision can also be inelastic, where the neutron is emitted at a lower energy and the remainder of the energy is emitted as a gamma ray of characteristic energy. See Figure 8.8.

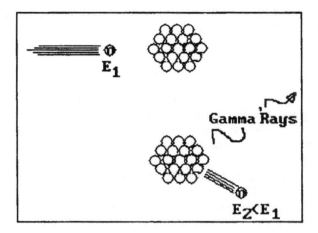

FIGURE 8.7
Inelastic scattering of a neutron where target nucleus absorbs a fast neutron and emits a slow one and gamma rays.

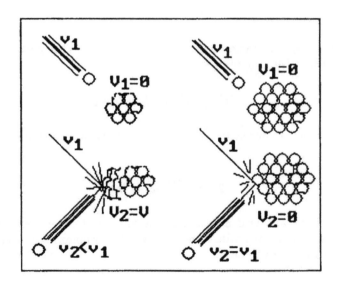

FIGURE 8.8
Elastic collisions of neutrons with nuclei involving some and negligible loss of energy.

At low energy, the neutron can be captured by the nucleus, which will then become unstable. It will eventually decay with the emission of gamma energy and/or particles.

The gamma rays given off by these processes have energies which are characteristic of the element nucleus involved and the type of reaction which took place. This is an important concept. It means that the originating elements and the types of reactions can be identified. The photon energy is a function of the wave frequency.

8.2.3 Protons and Similar Particles

A proton is a nuclear particle with a mass of approximately 1 AMU and a charge of +1 esu. It is the nucleus of a hydrogen atom, $^1H^{+1}$. Because of its charge and cross section, it is highly reactive. Therefore, few or no naturally occurring free protons. After capturing an orbital electron, the resulting hydrogen is stable.

An ionized hydrogen-2 isotope is the deuteron, $^2H^{+1}$. It has a mass of approximately 2 AMU and a charge of 1 esu. It is a nucleus of deuterium. It is a neutron and a proton combination. Deuterium is a stable element.

Another isotope of hydrogen is triton, ^3H. It has a mass of approximately 3 AMU and a charge of +1 esu. It is the nucleus of tritium, ^3H, or hydrogen-3. It is unstable (radioactive) and has a half life of 12.33 years. It decays with the emission of a beta particle:

$$2\left(^3H\right) \vee \left(^4He\right) + \left(^{-1}\beta\right) \qquad (8.13)$$

At high energy, a triton and a deuteron can react with the production of a neutron and 14 MeV of energy:

$$\left(^3H\right) + \left(^2H\right) \vee \left(^4He\right) + {}^0n + Q \qquad (8.14)$$

There can be a similar reaction producing a 4 MeV neutron between 2 deuterons:

$$2\left(^2H\right) \vee \left(^3He\right) + \left(^0n\right) \qquad (8.15)$$

These reactions are used in linear accelerators and other devices for the production of neutrons. Miniature accelerators are commonly used in geophysical logging equipment.

8.2.4 Alpha Particles

An alpha particle has a mass of approximately 4 AMU and a charge of +2 esu. It is the nucleus of a helium-4 atom, $^4He^{++}$. It has two protons and two neutrons in combination. Alpha particles are common reaction particles of the nucleus disintegration. They are highly reactive because of their mass and electric fields (large cross sections). They are not very penetrating because of their large cross section. ^4He results from a capture of two negative electrons, ^{-1}e. ^4He is a stable element.

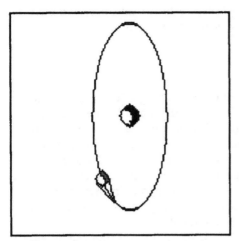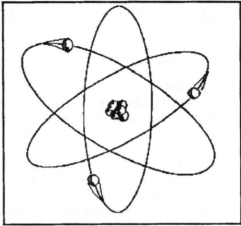

FIGURE 8.9
Bohr models of the hydrogen (left) and the lithium (right) atoms.

8.2.5   Atoms

The atomic model used in this text is the Bohr atom. It was proposed by Niels Bohr in 1913. It is a simple model by today's standards, but will suffice for our purposes (Figure 8.9). In the Bohr model, each atom has a nucleus which contains one or more protons. Because of its charge, each proton results in one orbital electron in a complete atom. Thus, indirectly, the protons determine the valence and, thus, chemical characteristics of the atom. The number of protons is the atomic number, Z, of the element. The proton can be considered to be the nucleus of a hydrogen atom, $_1^1H$.

In addition to the protons, the elemental nucleus contains one or more neutrons (except for hydrogen-1, $_1^1H$, which has none). Because they have no charge, the neutrons do not affect the valence of the atom; the chemical properties.

The combinations of a given number of protons with the various numbers of neutrons are the nuclides or isotopes of that element. The sum of the masses of the nuclear elements is the atomic mass, A, of that element isotope. The ratio of the atomic number, Z, to the atomic mass, A, is the Z/A ratio of that nuclide. The first 20 atomic elements, except for $_1^1H$, have average Z/A ratios of approximately 0.5. These are the most common elements in sedimentary geological materials. A Z/A ratio of 0.5 indicates an equal number of protons and neutrons in the nucleus. Since an element will have one orbital electron for each nuclear proton, if the Z/A ratio is known, the number of orbital electron will be a function of the mass of the element atom.

Just as a proton is a hydrogen-1 nucleus, an alpha particle is a helium-4, $_2^4He$, nucleus. An alpha particle, α, has an atomic mass (A) of 4 and an atomic number (Z) of 2. Its charge is +2 electrostatic units, esu. Table 8.1 lists some of the common elementary, low-mass particles.

8.2.6   Photons

A photon is a single, irreducible amount of electromagnetic energy. It can be in any frequency range, although it is usually used to describe the higher frequencies, such as visible light, X-rays, gamma rays, etc. It has zero mass. Its energy is a function of its frequency. Lower frequencies (longer wave lengths) indicate lower energies. It moves at the speed of light in any medium. It can be refracted and reflected like any wave. The only real difference between gamma rays and X-rays is in their origins. X-rays originate from orbital electron reactions and gamma photons from nuclear reactions. Once created, they are identical.

Photons travel discrete paths like particles. They act like particles, because of their energy. They have momentum. They can be reflected like particles where their angles of reflection are equal to their angles of incidence. They scatter by elastic or inelastic interaction with orbital electrons, like particles. A photon will be scattered at a lower energy.

Photons can undergo several types of scattering reactions, three of which are important to geophysical logging. The most important is elastic or Compton scattering. This takes place between an incident high energy photon and an orbital electron. The incident photon ejects the electron from its orbit by giving some of its energy to the electron. The photon scatters at a predictable angle and at a lower energy (lower frequency; longer wave length). The electron becomes a free photoelectron. This is shown in Figures 8-10, 8-11 and 8-12.

Lower energy photons can also interact inelastically with the orbital electrons and impart all of their energy to the electron, especially if the atom is large. The photon will cease to exist and the electron becomes a photoelectron. See Figure 8.13.

A third important type of reaction is pair production. In this reaction, the incident photon, if it has sufficient energy (>1.02 MeV), will interact with the nucleus and its energy will be converted to a negative and a positive electron pair. The negative electron becomes a free electron. The positron has a short life because it will react easily with any nearby electron and both will be annihilated, with the release of two photons of 511 keV of energy, the mass conversion of the two particles. This process is illustrated in Figure 8.14.

Gamma rays, γ, are electromagnetic energy. They originate from nuclear reactions by capture of another particle (inelastic reaction) by the nucleus (see Figure 8.7), the radioactive decay of the nucleus, or elastic scattering of a particle by the nucleus (see Figure 8.10). Particles may also interact with each other. The particles may be neutrons (Figure 8.7 and 8-8), alpha

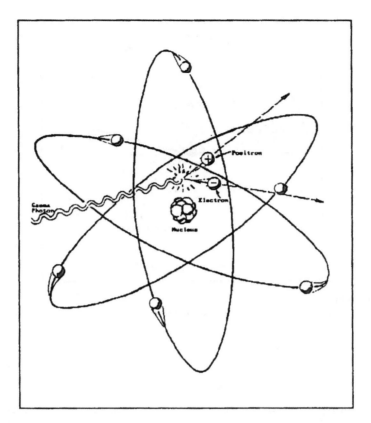

FIGURE 8.10
Compton scattering.

particles (Figure 8.15), beta particles (negative electrons) (Figure 8.4), or positrons (positive electrons). See Table 8.1. The resulting electromagnetic radiation (heat, radio waves, light, gamma radiation) represents the energy lost by the photon, the particle, and/or the nucleus in the reaction. The energy may also come from loss of some of the particle mass or the nucleus mass. It may also originate from disruption of forces, such as binding energy. One such reaction is shown in Figure 8.5.

Gamma rays may also originate in the spontaneous break-up or decay of unstable nuclei. When this happens, one or more particles and one or more gamma photons are emitted from the nucleus. Figure 8.16 shows the probable distribution of gamma ray contributing nuclei detected by a downhole gamma ray probe.

8.2.6.1 Gamma-Photon Scattering Interactions

Gamma rays interact with atoms in three ways:

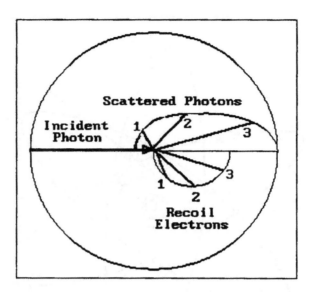

FIGURE 8.11
Scattered quanta and recoil electrons in Compton scattering.

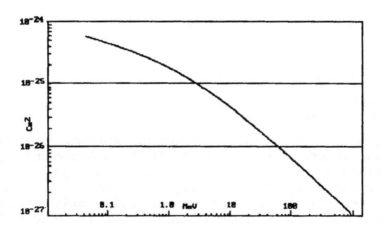

FIGURE 8.12
The Compton cross-section, per electron.

1. Photoelectric absorption results in the immediate ejection of an electron (the photoelectron). The photoelectron has absorbed all of the energy of the incident gamma photon. This reaction is dominant at energy levels, of the gamma photon, of ≤50 kev. The electron is usually from the K shell of the atom and will have an energy of:

$$E_e = E_\gamma - E_{binding} \qquad (8.16)$$

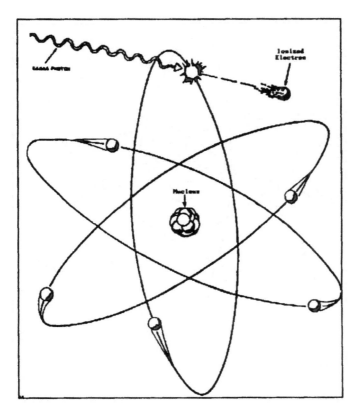

FIGURE 8.13
The photoelectric effect.

The gamma ray is annihilated in the photoelectric process. At energies above the K edge, the photoelectric absorption cross section, s_e, per electron increases as a function of $Z^{3.6}$. See Figure 8.13.

2. Compton scattering occurs at gamma ray energies, E∆, in the range of

$$-0.1\,MeV < E_\gamma < \sim 10\,MeV \qquad (8.17)$$

The electron energy relationship is the same in this reaction as in the photoelectric reaction above, but, since the photon has a higher energy, it will continue to exist after the reaction, but have a lower energy. Since the wavelength of the photon, λ_γ, is very short, the scattering takes place from individual electrons (Lapp and Andrews, 1948). The scattering kinetics are determined by the relativistic conservation of energy and linear momentum. In

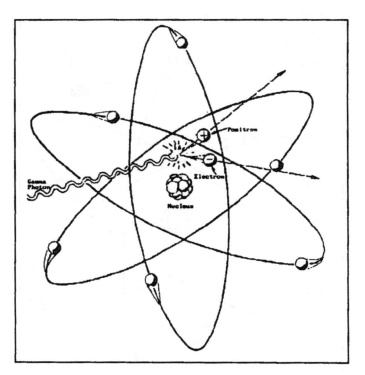

FIGURE 8.14
Pair production.

the Compton energy range, the electron cross section, σ_e, is independent of the atomic number, Z, for the elements commonly encountered (1 to 20). See Figure 8.2b.

3. Pair production is the creation of an electron-positron pair by interaction of the atom with the photon. This reaction will occur only when

$$E_\gamma \geq 2m_0c^2 = 1.02\,MeV \qquad (8.18)$$

where m_0c^2 = the rest mass of the electron and the positron. Again, the photon is annihilated in the reaction. The electron and the positron share equally the excess energy,

$$E_{pp} = E_\gamma - 2m_0\,c^2 \qquad (8.19)$$

In the materials (elements) where we will make density measurements, this reaction is significant only when $E_\gamma \geq 10$ MeV. In the pair production range, σ_e is approximately linear with Z.

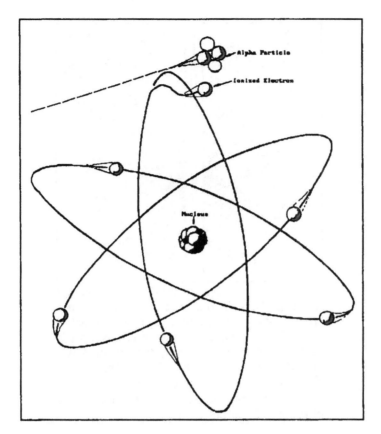

FIGURE 8.15
Alpha particle interaction.

Therefore, if gamma energies are restricted to the Compton region, at some distance from the source the gamma flux must be only a function of the number of electrons per unit volume. By the same reasoning, the gamma flux in the photoelectric region is very sensitive to Z.

8.2.6.2   Sources and the Scattering Process

At this time, the scattered gamma ray methods all use isotopic sources to put moderately high energy gamma rays into the formation material. The introduced gamma ray intensity is usually substantially above that of the natural gamma ray level. The amount of gamma radiation received at the detector is a function of the number of electrons per unit volume of formation material. There are no commercial electrical gamma ray sources available in logging today. Investigations are promising, but the work has a low priority. Table 8.2 shows some of the common gamma emitting isotopes in present use.

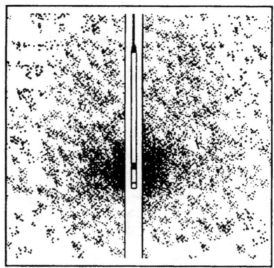

An illustration of the sample volume concept as a probability distribution. Although the chance of a given gamma ray being detected falls off rapidly with the distance between the ewmitter and the detector, that chance will nevertheless be finite. Depending upon gamma ray energy, rock density, and other parameters, this diagram represents an area of roughly 2 meters on a side.

FIGURE 8.16
Distribution of detected gamma emitters. (Courtesy of International Atomic Energy Agency.)

The process of density logging uses medium energy gamma rays from isotopic sources (660 kev from ^{137}Cs or 1.17 MeV from ^{60}Co). The coal tool (which does not measure density quantitatively) uses a source of ^{241}Am. The most commonly used source in petroleum density work is ^{137}Cs whose half life is 33 years. ^{60}Co is also used and its half life is about 5 years. ^{226}Ra is occasionally used because the licensing problems used to be easy (no longer so). ^{226}Ra has a half life of 1700 years, but its energy range is from 190 kev to 2.2 Mev. This wide range of energies causes calibration and linearity problems; besides, shields for the 2.2 MeV gamma photons are heavy and bulky.

For density measurements, the source energy must be high enough to favor Compton scattering and low enough to ease shielding problems and reduce pair production. Figure 8.17 shows the cross section to gamma ray photons, as a function of atomic number, Z, for two energy levels. For example, a ^{60}Co source, with gamma emissions of 1.17 MeV energy would

TABLE 8.2

Commonly-Used Gamma Ray Source Materials

Material	Symbol	Half-Life	Radiation (rhm/curie)	HVT Lead	Radiation (MeV)
Iodine 131	$_{53}I^{131}$	8.05 d	0.21	0.14"	β0.25(2.8%),0.335(9.3%) 0.608(87.2%),0.815(0.7%) γ0.080(2.2%),0.163(0.7%) 0.284(5.3%),0.364(80%) 0.637(9%),0.722(3%)
Iridium 192 $_{77}Ir192$	74.5 d	0.50	0.19" 0.537 γ0.316 1.159		β0.1(1.5%),0.257(7%) (41%,0.673(48%) (32%),numerous others from 0.136
Cobalt 60	$_{27}Co60$	5.27 y	1.24	0.49	β0.306 γ1.17(100%),1.33(100%)
Cesium 137$_{55}Cs^{137}$	30+/−3 y				
Barium 137$_{56}Ba^{137}$	2.6 m	0.32		0.25	β0.51(92%),1.17(8%) γ0.662
Radium 226	$_{88}Ra^{226}$	1620 y	0.825	0.56	α4.8(94%),4.6(6%) γ0.19(With high contamination from daughters, from 0.2 to 2.2)
Krypton 85	$_{36}Kr^{85}$	10.27 y	0.30	0.004	β0.15(65%),695(99+%) γ0.54(65%)

require a protective shield 2.9 times heavier, for the same reduction of intensity, than a ^{137}Cs source, with an emission energy of 660 kev. Pair production occurs at gamma energies greater than 1.02 MeV. Thus, pair production does not occur with a ^{137}Cs source. It only occurs to a small degree with ^{60}Co.

The source energy must also be high enough to enhance the Compton scattering process and minimize the relative amount of photoelectric (PE) reactions. The PE reaction is described by the Einstein equation:

$$E_{e,pe} = h\nu = E_w + \frac{1}{2}mv^2 \qquad (8.20)$$

where mv^2 represents the kinetic energy of the ejected electron or photo-electron

h = Planck's Constant and is equal to 6.6234×10^{-27} erg seconds

$h\nu$ = the total energy of the incident photon

E_w = the energy required to dislodge the electron from its atomic orbit (binding energy)

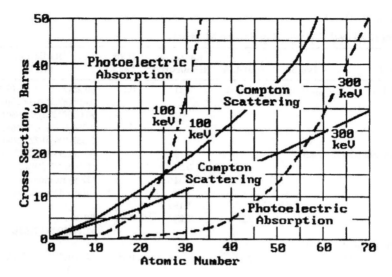

FIGURE 8.17
Relationship between Compton scattering and photoelectric effect.

Since the photon disappears in the PE reaction, it is of no use in the determination of the electron density. In addition, the PE reaction is quite nonlinear with respect to the atomic number or atomic weight. Thus, this reaction interferes with the determination of bulk density. It is, however, a sensitive indicator of rock type. The cross section, σk, of the K-orbit electrons, for energies up to 0.5 MeV, is,

$$\sigma_k = C \frac{Z^5}{E_w^{7/2}} cm^2 \tag{8.21}$$

where C is a constant. Figure 8.18 shows the relationship of Equation 8.21 for several elements.

Tittman (Tittman, 1986) also evaluates the function, τ, which is a photoelectric absorption cross section:

$$\tau = C \frac{Z^{3.6}}{E^{3.15}} \tag{8.22}$$

8.2.6.3 Compton Interactions

The relationship of the energy of the scattered gamma photon, E_γ with respect to the initial energy, $E_{\gamma 0}$ is:

$$\frac{E_\gamma}{E_{\gamma 0}} = \left[1 + \frac{E_{\gamma 0}}{m_0 c^2}(1 - \cos\theta)\right]^{-1} \tag{8.23}$$

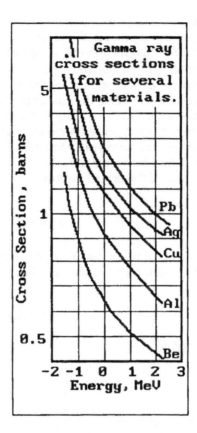

FIGURE 8.18
Approximate gamma ray cross
sections for several metals.

where θ is the angle of scatter. From Equation 8.23, it is evident that the energy after scattering is *not* independent of the source energy (before scattering).

Since m_0c^2 is a constant, the energy of the photon, E_γ depends upon the energy of the incident photon, E_{γ_0}, and its scattering angle, θ. Evaluating Equation 8.23,

$$\langle \Delta E_\gamma \rangle = \left[\frac{1}{\sigma_c}\left(E_{\gamma_0}\right)\right]\int_0^{E_{\gamma_0}}\left(E_{\gamma_0} - E_\gamma\right)\sigma_c\left(E_{\gamma_0} \to E_\gamma\right)dE \qquad (8.24)$$

we obtain an expression for the mean loss of energy, as a function of the photon energy before scattering, where $\sigma_c(E_{\gamma_0})$ is the total Compton cross section for scattering at E' and $\sigma_c(E_{\gamma_0} \to E_\gamma)$ is the partial cross section for scattering from E_{γ_0} into a unit energy interval at E_γ. Figure 8.12 shows the mean relative energy loss with decreasing gamma photon energy.

Tittman (Tittman, 1986) points out that the Compton cross section varies smoothly with energy and follows the Klein-Nishina relationship (Heitler, 1954). Since the relationship between the number of electrons, the mass of the nucleus, and the bulk density is known or can be determined, the scattered gamma ray flux is, therefore, a direct measure of the bulk density, ρ_b. He also says that the mixing of element components is intrinsically linear by volume. Thus, a material need not be monatomic for determination of ρ_b.

Compton scattering involves the orbital electrons of the target atom. At energies above 200keV, the amount of Compton scattering is almost linear with the atomic number of the elements 2 to 20. These are the chief constituents of the common sedimentary materials. Therefore, it is desirable to use high energy sources (0.5 to 1 MeV) and detect the Compton scattering for the determination of density.

The mass of the orbital electron, m_e, is:

$$m_e = m_o\left(1-\beta^2\right)^{-\frac{1}{2}} \tag{8.25}$$

where m_o is the rest mass of the electron and β is the ratio of its velocity, v, to that of light, c. Its momentum, M_e, is:

$$M_e = m_o v\left(1-\beta^2\right)^{-\frac{1}{2}} \tag{8.26}$$

Its energy, E_e, is

$$E_e = \left(m_e c^2 - m_o c^2\right)^{\frac{1}{2}} \tag{8.27}$$

These same quantities for the photon are its mass, m_p,

$$m_p = \frac{hv}{c^2} \tag{8.28}$$

its momentum, M_p,

$$M_p = \frac{E_p}{c} = \frac{h}{\gamma} = \frac{hv}{c} = mc \tag{8.29}$$

and its energy, E_p,

$$E_p = hv \tag{8.30}$$

In the scattering process (refer to Figure 8.11) the energy is conserved. Therefore, the energy, E_p, of the incident photon is:

$$E_p = hv' + mc^2 - m_o c^2 \tag{8.31}$$

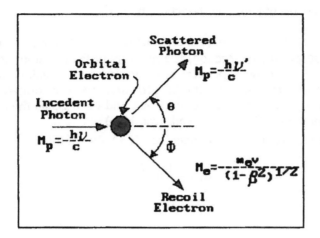

FIGURE 8.19
Photon scattering.

The parallel (to the path of the incident photon) component of the momentum is

$$\frac{h\nu}{c} = m_o v\left(1-\beta^2\right)\cos\phi = \frac{h\nu'}{c}\cos\phi \qquad (8.32)$$

and the perpendicular component is:

$$m_o v\left(1-\beta^2\right)^{-\frac{1}{2}}\sin\phi = \frac{h'}{c}\cos\phi \qquad (8.33)$$

The angle at which the electron scatters, I, is related to the angle at which the photon scatters, T, and the incident photon energy, E_p:

$$\tan\phi = \frac{0.51\cot\left(\dfrac{\Theta}{2}\right)}{E_p + 0.51} \qquad (8.34)$$

The pattern of scattering from a beam of gamma photons in Compton scattering is approximately shown in Figures 8-2 and 8-19. You can see that a substantial number of gamma photons are scattered away from the path of the incident beam. These scattering events are a function of the number of orbital electrons in the target elements. The number of electrons is proportional to the average atomic number of the formation material.

The measures of gamma interaction, then are,

1. The cross section, σ, (τ is often used to denote the photoelectric absorption cross section) is a measure of the apparent area of a target particle. A parallel beam incident on a slab containing many particles, will exit with a fraction, $e^{-n\sigma t}$, transmitted; where n is the number of target particles, and t is the slab thickness in centimeters (Lapp and Andrews, 1948).

2. Linear absorption coefficient, μ_l. The value of μ is,

$$\mu_l = -\frac{1}{x}\ln\frac{I}{I_o} = n\sigma \qquad (8.35)$$

where

x = the distance traversed by the beam, in centimeters

I = the intensity of the emerging beam

I_0 = the original intensity of the beam

The linear coefficient is a function of the macroscopic cross section of neutrons.

3. The mass absorption coefficient, μ_m,

$$\mu_m = \frac{\mu_l}{\rho} \qquad (8.36)$$

The mass coefficient is useful in the Compton region where the density, ρ, is the only parameter characterizing the target material. The symbol, μ is sometimes used for this coefficient. The dimensions or the context will indicate which parameter is meant.

4. Tittman (Tittman, 1986) shows the detail of the gamma ray transport mechanisms.

The spectrum, within the Compton range, resulting from scattering from a 0.5 MeV source shows a smooth response curve with no sensitivity to material type. (Tittman, 1986).

8.3   Geophysical Logging Considerations

The two particles of most interest in radioactivity exploration methods are the gamma ray photon, γ, and the neutron, n. Natural gamma ray detection is a very important part of geophysical logging. The total gamma emission (an indicator of the shale content, among other things) and the spectrographic recordings of natural gamma rays (an indicator of the elemental content of the formation material) are both used. Absorption and scattered gamma ray techniques are used extensively to determine density, porosity, and average atomic number of the formation zones.

Neutron logging is used in many varieties. The scattered neutrons are detected for porosity determination and uranium identification. They are used with scattered gamma rays to determine many of the characteristics of the formation materials. The gamma emission and the neutron emission, due to bombardment of materials by and capture of neutrons is becoming more important every day for identifying formation elemental content.

In addition to the gamma and neutron logging, some techniques use beta detection and X-radiation scattering, emission, and refraction. These are mostly surface and laboratory techniques, but some are used downhole, also. Deuterium and tritium reactions are used in neutron generators. Proton and neutron evaluation are important in density logging and neutron logging. The alpha and beta particles become important in understanding many of the logging and geological techniques and processes.

8.4   Health Physics

Generally, health physics is not within the scope of this text. There are a few things, however, that the user of radioactivity logging equipment should know.

Usually, the sources used with logging equipment are small and relatively safe (they are energy sources, however, and must be treated with respect). They are sealed in double and triple containers which are built to withstand high temperatures, high pressures and corrosive environments. They are carried in adequately shielded containers, until they are needed for use. Some of the newer types can be turned off or otherwise inactivated when not in use.

When the sources are in use, the exposure is generally brief. Simple rules have been set up which, if they are correctly followed, will insure that no one, not even the operator, will receive more than a very minimal exposure. Casual observers of logging operations run no risk from the logging radioactivity.

Glossary

Symbols, Abbreviations, Subscripts, and Superscripts

A	Absolute; area; atomic weight (mass)
a	Apparent (partially or wholly uncorrected); tortuosity factor
Ag	Silver
α	Alpha particle; proton
Å	Ångstrom unit
AMU, amu	Atomic mass unit
~	Approximately
$\approx$, $\cong$	Approximately equal to
atm	Atmosphere of pressure
Au	Gold
API	American Petroleum Institute; unit approved by the API
APIg	Arbitrary gamma ray logging units approved by the API
APIn	Arbitrary neutron logging units approved by the API
b	Bulk
Ba	Barium
β	Beta particle; electron; ratio v:c
BIB	Bibliography
BHT	Bottom hole temperature
C	Carbon; centigrade; celsius; conductivity (electrical)
Ca	Calcium
c	Electrical conductance; core
Cal	Caliper
cc	Cubic centimeter
CEC, C.E.C.	Cation exchange capacity
Ci	Curie
Cl	Chlorine
cl	Clay
cm	Centimeter
corr	Corrected
Csg, csg	Casing
D	Depth; detector; bit size (diameter)
d	Diameter; darcy; deep; mathematical differential; dry; diminution; rank difference
Δ	Difference
diff	Difference
dol	Dolomite

E	Voltage; volts; voltage source
e	Effective; electron
EMU, emu	Electromagnetic unit
est	Estimate
ESU, esu	Electrostatic unit
eV	Electron volt
F	Fahrenheit
FE	Formation evaluation
fl	Fluid; flow line
ft	Foot
f, form	Formation
F_r	Formation resistivity factor
G	Geometrical factor
g	Gamma ray; gram; gas
g/cc, g/cm^3	Density units, grams per cubic centimeter
Δ	Gamma ray photon; gamma ray
H	Hydrogen
h	Hydrocarbon; borehole; Planck's constant
He	Helium
I	Electrical current
i	Invaded; invasion; invaded zone
IL	Induction log
I_r	Saturation index; invaded
ir	Irreducible
K	Potassium; relative permeability
k	Permeability; kilo- ($\times 10^3$)
M	Lambda; wave length; disintegration constant; probability constant
L,l	Length; linear distance; lag
Log, log	Mathematical logarithm (base 10)
Ln, ln	Mathematical logarithm (base e)
ls	Limestone
M	Mass; total rock cementation exponent, mega-($\times 10^6$)
m	Cementation exponent; meter; medium; mud
mc	Mudcake subscript
mf	Mud filtrate subscript
Mg	Magnesium
ML	MicroLog
MLL	Microlaterolog
MOP	Moveable oil plot
MOS	Moveable oil saturation
MSFL	Microspherically focussed log
μ	Micro; micron; viscosity ($\times 10^{-6}$)
–	Negative; minus (mathematical)
MWD	Measurement while drilling
N	Neutron

n	Saturation exponent; neutron; number (mathematical)
Na	Sodium
ν	Frequency
O	Oxygen
o	100% water saturated; oil; degree (superscript); initial (subscript)
Ω	Ohm
Ωm	Ohmmeter(s); Ohms meter(s)^2 per meter
P	Pressure
p	Pipe
Pa	Protactinium
Pb	Lead
Φ	Angle of electron scatter
φ	Porosity
pe	Photoelectric
+	Positive; plus (mathematical)
ppb	Parts per billion
ppm	Parts per million
%	Percent; 1/100
∝	Proportional to
psi	Pounds per square inch
PSP	PseudoSP (affected by clay mineral)
φ	Porosity
Q,q	Quantity; Volume per unit time
qz, qtz	Quartz
R	Electrical resistivity
r	Electrical resistance; radius; radial distance; residual; resistivity
Ra	Radium
Ref	Reference; references
ρ	Density
Rn	Radon
S	Saturation; sulfur; source
s	Saturated; surrounding; sand; second
sh	Shale
Si	Silicon
σ	Micro cross section
Σ	Sum; mathematical summation; thermal neutron macro cross section
SP	Spontaneous (self) potential
SSP, ssp	Static SP value
Sym, sym	Symbol
T	Time
t	True; Unit (Interval) travel time; circulation time; total; tortuosity
t	Photoelectric cross section

TD	Total depth
Th	Thorium
Θ	Photon scattering angle
θ	**Angle**
V	Volume
v	Velocity
w	Water; wet
W	Weight
WC	Water cut
WGR	Water:gas ratio
WOR	Water:oil ratio
xo	Pertaining to the invaded zone
Z	Atomic number
Z/A	Atomic number:atomic mass ratio

Bibliography

Albright, J.N. and Pearson, C.F., *Acoustic Transmissions as a Tool for Hydraulic Fracture Location*, JPE, Dallas, August 1982.

Alger, Robert P., and Harrison, Charles W., Improved fresh water assessment in sand aquifers utilizing geophysical well logs, *The Log Analyst*, Vol. 30, No. 1, SPWLA, Houston, Jan-Feb 1989.

Allaud, Louis and Martin, Maurice, *Schlumberger, The History of a Technique*, John Wiley and Sons, 1977.

Amyx, J.W., Bass, D.M., Whiting, R.L., *Petroleum Reservoir Engineering*, McGraw-Hill Book Co., Inc., New York, 1960.

Archie, Gerald E., The electrical resistivity log as an aid in determining some reservoir characteristics, *A.I.M.E.* Transactions, v. 146, pp. 54–61, 1942.

Bigelow, Edward L., *Fundamentals of DIPLOG Analysis*, Dresser Atlas, Dresser Industries, Houston, 1987.

Boatman, E. M., *An Experiment of Some Relative Permeability — Relative Electrical Conductivity Relationships*, Master's Thesis, Dept. of Pet. Eng., University of Texas, Austin, TX, June, 1961.

Brown, A. and Hussein, S., *Permeability from Well Logs, Shaybah Field, Saudi Arabia*, Transactions of the 18th S.P.W.L.A. Symposium, 1977.

Burke, J.A., Schmidt, A.W. and Campbell, R.L.,Jr., *The Litho Porosity Cross Plot, The Log Analyst*, Vol. X, No. 6, SPWLA, Houston, Dec 1969.

Cassel, Bruce, *Vertical Seismic Profiles — An Introduction*, Western Geophysical Company, Middlesex, U.K., 1984.

Calhoun, John C., Jr., *Fundamentals of Reservoir Engineering*, University of Oklahoma Press, Norman, 1955.

Carmen, P. C., *Flow of Gases Through Porous Media*, Academic Press, Inc., New York, N. Y., 1956.

Chapellier, Dominique, *Diagraphies Appliquees a L'hydrologie*, Lavoisier TEC & DOC, Paris, 1987.

Clark, Isobel, *Practical Geostatistics*, Applied Science Publishers, London, 1979.

Clavier, C. and Rust, D.H., Mid Plot: A New Lithology Technique, *The Log Analyst*, SPWLA, Vol. XVII, No. 6, Houston, Nov–Dec 1976.

Cork, James M., *Radioactivity and Nuclear Physics*, D. Van Nostrand Co., Inc., Princeton, N.J., 1957.

Crain, E.R., *The Log Analysis Handbook*, PennWell Publishing Co., Tulsa, 1986.

Dakhnov, V.N., (Keller, G.V., ed.), Geophysical Well Logging, *Colorado School of Mines Quarterly*, Vol. 57, No.2, Golden, Colorado, 1962.

Darcy, H., *Les Fontaines Publiques De La Ville De Dijon*, Victor Dalmont, Paris, France, 1856.

Degolyer, E., Notes on the early history of applied geophysics in the petroleum industry, The *Journal of the Society of Petroleum Geophysicists*, Division of Geophysics, A.A.P.G., Vol. VI, No. 1, July 1935.

Dewan, J.T., *Essentials of Modern Open Hole Log Interpretation*, PennWell Books, Tulsa, 1983.

Desbrandes, Robert, *Encyclopedia of Well Logging*, Gulf Publishing Co., Houston, 1985.

Doll, Henri G., *The SP Log in Shaly Sands*, AIME Paper T.P. 2912, 1949.

Doveton, John H., *Log Analysis of Subsurface Geology*, John H. Wiley and Sons, 1986

Dresser Atlas, *Well Logging and Interpretation Techniques*, Dresser Industries, Houston, 1982.

Dresser Atlas, *Log Interpretation Charts*, 1985

Englehart, W. V. and Pitter, H., *Uber Die Zusmamenhangen Zwischen Porositat, Permeabilitat, Und Korgrobe Bei Sanden Und Sandstein*, Heidel. Beitr. Petrogr., 2, 1951.

Fertl, W.H., *Abnormal Formation Pressures*, Elsevier Scientific Publishing Company, Amsterdam/New York, 1976.

Fertl, W. H., and Vercellino, W. C., Predict Water Cut from Well Logs, *Oil and Gas Journal*, June 19, 1978.

Fertl, W.H., Wichmann, P.A., How to Determine Static BHT from Well Log Data, *World Oil*, January 1977.

Frasier, D.C., Keevil, N.B., Jr. and Ward, S.H., Conductivity Spectra of Rocks from the Craigmont Ore Environment, *Geophysics*, V. 29, no. 5, pp. 832–847, 1964.

Gearhart Industries, Formation Evaluation Data Handbook, Fort Worth, Texas, 1982.

Geyer, R.L. and Myung, J.I., *The 3-D Velocity Log; A Tool for In-Situ Determination of the Elastic Moduli of Rocks*, Seismograph Service Corporation, Tulsa, 1970.

Glossary of Terms and Expressions Used in Well Logging; Second Edition, Society of Professional Well Log Analysts, Houston, 1984.

Gondouin, M., Tixier, M.P., and Simard, G.L., An Experimental Study on the Influence of the Chemical Composition of Electrolytes on the SP Curve, *Journal of Petroleum Technology*, Feb. 1957.

Green, William, R., *Computer-Aided Data Analysis*, John Wiley & Sons, New York, 1985.

Hallenburg, J.K., *HP41C Formation Evaluation Programs*, PennWell Books, Tulsa, 1984.

Hallenburg, J.K., *Logcomp, Petroleum Formation Evaluation Programs*, PennWell Books, Tulsa, 1985.

Helander, D.P., *Fundamentals of Formation Evaluation*, OGCI Publications, Tulsa, 1983.

Holt, Owen R., *Relating Diplogs to Practical Geology*, Dresser Atlas, Dresser Industries, Houston, 1980.

Jones, P. J., Production Engineering and Reservoir Mechanics (Oil, Condensate, and Natural Gas), *Oil and Gas Journal*, 1945.

Keller, G.V. and Frischknecht, F.C., *Electrical Methods in Geophysical Prospecting*, Pergamon Press, Oxford, 1966.

Keller, G.V., Electrical Prospecting for Oil, *Colorado School of Mines Quarterly*, Vol. 63, No. 2, April 1968.

Koerperich, E.A., Shear Wave Velocities Determined from Long and Short Spaced Borehole Acoustic Devices, *JPE*, October 1980, Dallas.

Koerperich, E.A., Investigation of Acoustic Boundary Waves and Interfering Patterns as Techniques for Detecting Fractures, *JPE*, Dallas, August 1978.

Lapp, R.E., and Andrews, H.L., *Nuclear Radiation Physics*, Prentice-Hall, Inc., New York, 1949.

Larinov V.V., *Borehole Radiometry*, Nedra, Moscow, 1969.

LeRoy, L.W. and LeRoy, D.D., *Subsurface Geology*, Colorado School of Mines, Golden, Colorado, 1977.

Lynch, Edward J., *Formation Evaluation*, Harper & Row, New York, 1962.

Meehan, D.N. and Vogel, E.L., *HP41C Reservoir Engineering Manual*, PennWell Books, Tulsa, 1982.

Morris, R. L. and Biggs, W. P., *Using Log-Derived Values of Water Saturation and Porosity*, Transactions of the 1967 S.P.W.L.A. Symposium, 1967.

Myung, J.I. and Helander, D.P., *Correlation of Elastic Moduli Dynamically Measured by In-Situ and Laboratory Techniques*, 13th Annual Logging Symposium, S.P.W.L.A., Tulsa, 1972.

Myung, J.L. and Henthorne, J., *Elastic Property Evaluation of Roof Rocks with 3-D Velocity Logs*, Solution Mining Research Institute, Atlanta, 1971.

NL Baroid/NL Industries, Inc., *Manual of Drilling Fluids Technology, The History and Functions of Drilling Mud*, Vol 1, Section 1, 1979.

Oliver, D.W., Frost, E., Fertl, W.H., *Carbon/Oxygen Log*, Dresser Atlas, Dresser Industries, Houston, 1981.

Overton, H.L., Lipson, L.B., A Correlation of Electrical Properties of Drilling Fluids with Solid Content, *A.I.M.E.*, 213:333–336, 1958.

Pirson, S.J., 1935, Effect of Anisotropy on Apparent Resistivity Curves, *Bull. A.A.P.G.*, v. 19, no. 1, pp. 37–57.

Raymer, L.L. and Biggs, W., *Matrix Characteristics Defined by Porosity Computations*, Schlumberger Well Services, c1970.

Recommended Practice for Determining Permeability of Porous Media, American Petroleum Institute, APR RP No. 27, Sept. 1952.

Schlumberger, A.G., *The Schlumberger Adventure*, Arco Publishing, Inc., New York, 1982.

Log Interpretation Principles/Applications, Schlumberger Educational Services, Houston, 1987.

Schlumberger Well Services, Inc., *Log Interpretation Charts*, 1986.

Sharma, P.V., *Geophysical Methods in Geology*, Elsevier, Amsterdam, 1986.

Sheriff, R.E., *Encyclopedic Dictionary of Exploration Geophysics*, Society of Exploration Geophysics, Tulsa, 1973.

Society of Professional Well Log Analysts, *The Art of Ancient Log Analysis*, 1979.

Timur, A., An Investigation of Permeability, Porosity, and Residual Water Saturation Relationships for Sandstone Reservoirs, *The Log Analyst*, July–August, 1968

Tittman, Jay, *Geophysical Well Logging*, Academic Press, Inc., Orlando, Florida, 1986.

Vennard, John K., *Elementary Fluid Mechanics*, John Wiley & Sons, New Yorl, 1961.

Weast, Robert C. (ed), *CRC Handbook of Chemistry and Physics*, 61st Ed., CRC Press, Inc., Boca Raton, Florida, 1981.

Winsauer, W.O., Shearin, H.M.,Jr., Masson, P.H., and Williams, H., Resistivity of Brine Saturated Sands in Relation to Pore Geometry, *American Association of Petroleum Geology Bulletin*, V 36, no. 2, 1952, pp. 253–277.

Zemenek, J., *Low-Resistivity Hydrocarbon-Bearing Sand Reservoirs*, SPE Paper 15713, Dallas, 1987.

Index

A

Abnormal pressures, *see* Overpressure, 1
Acoustic, 2, 14, 111
   stress, 1
   travel time, t, 30, 32, 34, 35, 55–53
   velocity, 34, 52–53
Activity, *see* Ions
Airborne, *see* Gamma ray, magnetic
Alpha, *see* Particle
American Petroleum Institute, API, 32
AMU, Amu, atomic mass unit, *see* Atom
Ångstrom unit, 146
Apparent resistivity, *see* Resistivity
Archie, 2, 131, 134, 135
Arkosic, *see* Formation
Atom
   AMU, Atomic mass unit, 146, 148, 149,
      150, 152
   atomic number, Z, 154
   Z/A, *see* Density

B

Bed, 135
   bed thickness, 1
Beta, *see* Particles
Borehole, 12, 83–101
   bridging, 96–97
   cased hole, 83,100–101
   caving, 96–97
   centrallizers, 101
   ledges, 97
   rugosity, 97
   tool position, 98–101
   tubing, 101
   water entry, loss, 33
Boyle's law, *see* Core(s), Boyles' law
Bridge, 89
Bridging, *see* Borehole
Brine, *see* Mud salinity
Bulk density, *see* Density

C

Cable tool, *see* Drilling cable tool
Calibration, 3, 103, 112

Caliper, 96–97
Carbonate, *see* Formation
Cased holes, *see* Borehole
Cation exchange capacity, CEC,
     see Spontaneous potential
Caving, *see* Borehole caving
Cement(ing), *see* Borehole
Charge, 148, 152, 154
Chinese, *see* Drilling Chinese
Chlorine log, *see* Neutron
Circulation, *see* Drilling circulation
Clay, shale, *see* Formation
Coal, *see* Formation
Compressional wave, Pressure wave,
     P-wave *see* Acoustic
Compton scattering, *see* Radioactivity
     Compton scattering
Conductance, conduction, *see* Resistance
Core(s), 9, 10, 66, 69, 83, 86, 103–119,
     128
   Boyles' law, 115
   core analysis, 4, 9, 60, 66, 86, 103, 106,
     109–119, 127
   core barrel, 104–105
   coring, 84, 98
   core-gamma, 4, 7
   core permeability, *see also* Permeability,
     116–119
   core porosity, 60–61, 115
   electrical, 103, 105
   Dean Stark, 112, 113
   density, *see also* Density,
     115–116
   drop coring, 108
   extractor, 112–116
   log, 8
   retorting, 114–116
   quality, 109–111, 118–119
   Soxhlet, 112, 114
   volume, 115, 116
   Washburn-Bunting, 115
Correction, 3, 98, 99
Cross-plot, 2, 69, 73, 78
   Hingle, 139–141
   Pickett, Rt vs ϕ, 138–139
Cuttings, *see* Drilling, drill cuttings, sample
     cuttings, density cuttings